Workbook 6
Troubleshooting and Failure Analysis

Dr. Medhat Kamel Bahr Khalil, Ph.D, CFPHS, CFPAI.
Director of Professional Education and Research Development,
Applied Technology Center, Milwaukee School of Engineering,
Milwaukee, WI, USA.

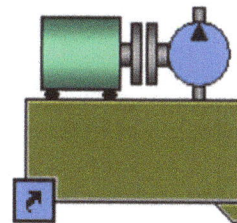

CompuDraulic LLC
www.CompuDraulic.com

CompuDraulic LLC

Workbook 6

Troubleshooting and Failure Analysis

ISBN: 978-0-9977634-7-8

Printed in the United States of America
First Published by June 2022
Revised by ----------------------

Disclaimer

It is always advisable to review the relevant standards and the recommendations from the system manufacturer. However, the content of this book provides guidelines based on the author's experience.

Any portion of information presented in this book couldn't be applicable for some applications due to various reasons. Since errors can occur in circuits, tables, and text, the publisher assumes no liability for the safe and/or satisfactory operation of any system designed based on the information in this book.

The publisher does not endorse or recommend any brand name product by including such brand name products in this book. Conversely the publisher does not disapprove any brand name product by not including such brand name in this book. The publisher obtained data from catalogs, literatures, and material from hydraulic components and systems manufacturers based on their permissions. The publisher welcomes additional data from other sources for future editions.

Workbook 6
Troubleshooting and Failure Analysis
Table of Contents

PREFACE

This Workbook is a complementary part to the textbook of the same title. This book is used as a workbook for students to take notes during the course delivery. It contains colored printout of the PowerPoint slides that are designed to present the course. Each chapter is followed by a number of review questions and assignments for homework.

Dr. Medhat Kamel Bahr Khalil

Chapter 1
Hydraulic Systems
Troubleshooting Logic Methodology

Objectives:

This chapter discusses the common methodologies applied for hydraulic system fault detection. This chapter introduces, in a step-by-step, the logic methodology for hydraulic system troubleshooting.

0

0

Brief Contents:

1.1- Fault Detection Methodology

1.2- Logic Fault Detection Procedure

1.3- General Check

1.4- Noisy Unit

1.5- Excessively Hot Unit

1

1

1.1- Fault Detection Methodology

Cost of Machine's Down time:
An hour saved in locating a problem →
$$$ saving in production.

Halfway to System Troubleshooting:
- Knowledge of components .
- Knowledge of machine history.

Guess and Miss Approach: *Random* component changing until the failed component is located.
- ❖ **Adv:** Easiness + no experience required
- ❖ **Disadv:** Good working components may be changed unnecessarily.

Logic Fault Detection Approach: Series of *logic* steps to find the cause of failure.
- ❖ **Adv:** Cost effective because only failed units will be repaired or replaced.
- ❖ **Disadv:** Experience required.

Video 184 (2 min)

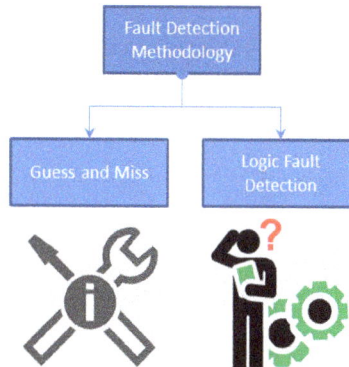

Fig. 1.1- Fault Detection Methodology

2

1.2- Logic Fault Detection Procedure

10 Analytical Steps for Logic Fault Detection

- ❖ **Step 1: Review Safety Instructions:**

Review safety instruction provided by:
- Machine manufacturer.
- Place in which the machine is used.
- Local municipality.

Fig. 1.2- Review Safety
Instructions before Start

In Volume 5 (Maintenance and Safety):
- BP-Safety-03: Safety of Hydraulic System Work Environment.
- BP-Safety-04: Safety of Hydraulic System Workspace.
- BP-Safety-05: Safe Startup of Hydraulic Systems.
- BP-Safety-06: Safe Operation of Hydraulic Systems.
- BP-Safety-07: Safe Servicing of Hydraulic Systems.

3

❖ **Step 2: Review Machine History:**   📹 Video 210 (0.5 min)

- The following information must be reviewed and understood:
  - Machine log.
  - Hydraulic circuit diagram.
  - Control circuit diagram.
  - Latest oil analysis report.

- The following questions must be answered:
  - What is the machine application?
  - Was the development of the fault gradual or sudden?
  - Has the fault occurred after oil changes?
  - Has the fault occurred after change of component change/adjustment?
  - Has the fault occurred previously, how frequently?
  - How does the fault affect the machine operation?
  - How does the fault affect the other connecting processes?

4

4

❖ **Step 3: Identify Main System Fault:**   📹 Video 182 (0.5 min)
- **T-System-01:** Fluid Aeration.
- **T-System-02:** Pump Cavitation.
- **T-System-03:** Excessive System Noise & Vibration.
- **T-System-04:** Excessive System Heat.
- **T-System-05:** Lack of Load Carrying Capacity.
- **T-System-06:** Faulty System Sequence.
- **T-System-07:** External Leakage.
- **T-System-08:** Troubleshooting Open Hydraulic Circuit.
- **T-System-09:** Troubleshooting Closed Hydraulic Circuit (Hydrostatic Transmission).
- **T-System-10:** Actuator Slow Performance.
- **T-System-11:** Actuator Fast Performance.
- **T-System-12:** Actuator Erratic Performance.
- **T-System-13:** Actuator Moves in Wrong Direction.
- **T-System-14:** Actuator Stops to Move.
- **T-System-15:** Actuator Load Drifts.
- **T-System-16:** Actuator Leaks.

Actions listed in each chart are ordered based on the common causes and the easiness of doing the action.

5

5

❖ **Step 4: Apply the System-Level Troubleshooting Chart:**
- Apply the ones that are assigned to the identified faults.
- Likely by applying this step the fault will be identified.

❖ **Step 5: List Suspicious Components:**
- If the fault isn't identified → list the suspicious components
- Start with the components that are easy to test and easy to access.

❖ **Step 6: Perform Preliminary Check on the Suspicious Components:**
- Inspect each suspicious component using the relevant inspection sheet
- Apply the following preliminary checks on each suspicious components:
  - **T-Unit-01:** General Check
  - **T-Unit-02:** Noisy Unit.
  - **T-Unit-03:** Excessively Hot Unit.

6

6

❖ **Step 7: Apply Detailed Check on the Suspicious Components:**
- **T-Seal-01:** Seal Troubleshooting.
- **T-Pump-01:** No Flow out of the Pump.
- **T-Pump-02:** Low Flow out of the Pump.
- **T-Pump-03:** Erratic Flow out of the Pump.
- **T-Pump-04:** Excessive Flow out of the Pump.
- **T-Pump-05:** No Pressure at the Pump Outlet.
- **T-Pump-06:** Low Pressure at the Pump Outlet.
- **T-Pump-07:** Erratic Pressure at the Pump Outlet.
- **T-Pump-08:** Excessive Pressure at the Pump Outlet.
- **T-Pump-09:** Leaking Pump.
- **T-Pump-10:** Excessive Pump Wear.
- **T-Pump-11:** Air Leaks into Pump.
- **T-Pump-12:** Excessive Pump Noise and Vibration.
- **T-Valve-01:** DCV Troubleshooting.
- **T-Valve-02:** FCV Troubleshooting.
- **T-Valve-03:** PCV Troubleshooting.
- **T-Valve-04:** EH Valve Troubleshooting.
- **T-Valve-05:** General Valve Troubleshooting.
- **T-Motor-01:** Motor Troubleshooting.
- **T-Cylinder-01:** Cylinder Troubleshooting.
- **T-Accumulator-01:** Accumulator Troubleshooting.
- **T-Reservoir-01:** Reservoir Troubleshooting.
- **T-Transmission Line-01:** Transmission Line Troubleshooting.
- **T-Heat Exchanger-01:** Heat Exchanger Troubleshooting.
- **T-Filter-01:** Filter Troubleshooting.

Video 185 (1 min)

Likely by then, the fault is identified. If not, detailed investigation is required. Review the relevant inspection sheet and apply the action in relevant check list.

7

7

❖ **Step 8: Fault Evaluation Decision of Repair or Replacement:**
- A decision must be made to repair or replace a faulty component based on evaluating the fault.

❖ **Step 9: Startup and Testing:**
- After removing the cause of fault by repair or replacement, a startup and testing step is required.
- Before starting-up, review safety precautions required for starting up a hydraulic-driven machine from Vol. 5.

❖ **Step 10: Future Considerations and Documentation:**
- Recommendations to assure that this fault won't occur again.
- Follow up on the system performance.

   Video 189 (1 min)

8

8

1.3- General Component Check

T-Unit-01-General Check	
model number or ordering code are incorrect?	▪ Replace the incorrect component based on the correct model number or ordering code.
Is the unit installed correctly?	▪ Follow the guidelines for proper installation of the component.
Does the unit receive a control signal?	▪ Check for proper control signal.
Is the unit adjustable?	▪ Check if the unit is properly adjusted.

Table 1.1

9

9

1.4- Noisy Unit

T-Unit-02-Noisy Unit	
Is the unit rotating?	▪ Check maximum speed. ▪ Check worn or sticking part
Pressure fluctuations in the return line?	▪ Check for restrictions in return line.
Is the unit coupled to a prime mover or coupled to a rotating mass?	▪ Check the coupling and alignment conditions.
Is the unit adequately supported to sub-plate or pipe works?	▪ Tighten the unit per the recommendation.
Mechanical noise?	Check: ▪ Loose, worn, or misaligned coupling. ▪ Loose set screw. ▪ Badly worn internal parts. ▪ Bearing failure.

Table 1.2

Video 211 (0.5 min)

10

10

1.5- Excessively Hot Unit

T-Unit-03-Excessively Hot Unit	
Is the unit experiencing internal leakage?	▪ Test the internal leakage and compare it with the allowable rate. ▪ Resolve the issue if found.
Is the unit undersized?	▪ Size the valve properly based on the flow rate.
The unit experiencing high internal friction or seizure?	▪ Check the unit and resolve the issue.
Is the unit rotating or cycling faster than the normal rate?	▪ Adjust the unit performance to the rated value.
Does the case drain restricted and is not cooled?	▪ Resolve restriction and consider cooling case drain.
Is viscosity too low causing lack of lubrication?	▪ Oil may be too thin either from wrong choice of oil or from thinning out at high temperature. Consequently, lack of lubrication causes system overheating and chain action continues.
Is ambient temperature too high?	▪ Consider proper ventilation around the unit.
Is fluid heavily contaminated by abrasives?	▪ Maintain the recommended cleanliness level.
Is the unit covered by dirt?	▪ Keep outside surfaces clean.

Table 1.3

11

11

Chapter 1 Reviews

1. Which of the following actions should be the step#1 when troubleshooting a hydraulic system?
    A. Reviewing the hydraulic circuit diagram.
    B. Reviewing safety instructions.
    C. Reviewing machine history.
    D. Reviewing the electrical circuit diagram.

2. The Guess and Miss troubleshooting approach has the following advantage?
    A. Fault is identified faster.
    B. Only component at fault is changed.
    C. It requires experienced personnel to perform it.
    D. It is an easy process that requires no experience.

3. The Logic Fault Detection troubleshooting approach has the following advantage?
    A. It is cost effective because only component at fault is changed.
    B. It requires experienced personnel to perform it.
    C. It is an easy process that requires no experience.
    D. It does not require to review the safety instructions at first.

4. When performing a general check on a hydraulic component, which of the following actions are among the checklist?
    A. Checking if the model number of the component is correct.
    B. Checking if the component is rotating at high speed.
    C. Check if the component is working at overpressure.
    D. Check if the component is working at high temperature.

5. When checking an excessively hot hydraulic component, which of the following actions are among the checklist?
    A. Checking if the model number of the component is correct.
    B. Checking if the component is rotating or cycling faster than normal.
    C. Check if the component receives electrical signal.
    D. Check if the component is adjustable.

Chapter 1 Assignment

Student Name: --- Student ID: ------------------

Date: -- Score: -----------------------

Question 1: When troubleshooting a hydraulic system, step 1 is to review safety instructions, list the sources of safety instructions you should consider?

Question 2: When troubleshooting a hydraulic system, step 2 is to review machine history, list what documents you should consider?

Chapter 2
Basic Troubleshooting Equipment

Objectives:

Servicing staff for hydraulic-driven machines should be aware of the troubleshooting equipment that are used in detecting faults of hydraulic systems. This chapter presents examples of troubleshooting equipment for hydraulic systems.

0

Brief Contents:

2.1- Snap-Check Pressure Gauge Test Kit
2.2- Hydrostatic Transmission Pressure Gauge Test Kit
2.3- Pressure/Leak Test Kit
2.4- Universal Flow Meter Test Kits
2.5- Portable Digital Hydraulic Multimeter
2.6- Adaptor Kit
2.7- Test Points and Pressure Measurement Hoses
2.8- Fluid Leakage Test Kit
2.9- Surface Temperature Thermometers
2.10- Vibration Indicators
2.11- Tachometers
2.12- Multimeters
2.13- Proportional Valve Tester
2.14- Servo Valve Tester

1

2.1- Snap-Check Pressure Gauge Test Kit

- 0 to 30″ Hg (0 to -1.0 bar) <u>vacuum gauge</u>
- 0 to 3000 PSI (0 to 207 bar) <u>pressure gauge</u>
- 0 to 6000 PSI (0 to 414 bar) <u>pressure gauge</u>
- Two (2) 60″ (1524 mm) <u>microbore hose assemblies</u>. Each hose assembly has one (1) male 'Snap-Check' test nipple and one (1) 'Snap-Check' test coupler on either end.
- Various sizes of 'Snap-Check' diagnostic test nipple.

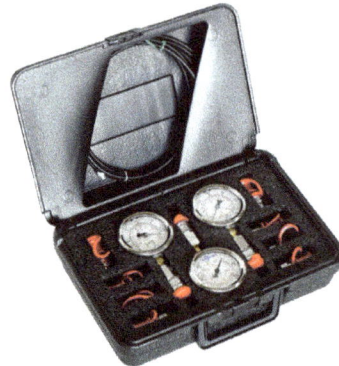

**Fig. 2.1- HC-TKC1-SC Test Kit
(Courtesy of Hydracheck)**

2

2

2.2- Hydrostatic Transmission Pressure Gauge Test Kit

- 0 to 100 PSI (0 to 6.9 bar) <u>pressure gauge</u>.
- 0 to 600 PSI (0 to 41.4 bar) <u>pressure gauge</u>.
- 0 to 6000 PSI (0 to 414 bar) <u>pressure gauge</u>.
- Three (3) 60″ (1524 mm) <u>microbore hose assemblies</u>.
- Three (3) microbore <u>hose unions</u> (connect hoses for additional length).
- One (1) 1/4″ NPT pressure test connector.
- One (1) 5/16″ SAE pressure test connector.
- Two (2) 7/16″ SAE pressure test connectors.
- One (1) 9/16″ SAE pressure test connector.

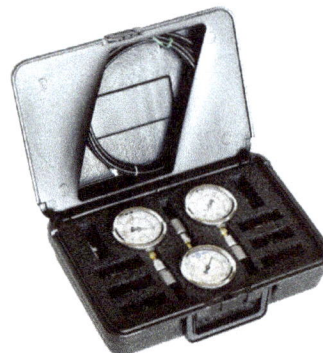

**Fig. 2.2- Hydrostatic Transmission
Pressure Gauge Test Kit
(Courtesy of Hydracheck)**

3

3

2.3- Pressure/Leak Test Kit

- 0 to 3000 PSI (207 bar) pressure gauge.
- 24" (607mm) hose assembly with STAUFF® Test 15 swivel nut.
- Pressure/leak test pump (3000 PSI {207 bar} maximum operating pressure).
- Infrared, non-contact, mini-thermometer, with laser sighting.
- NPT adaptors sizes 1/4" to 1-1/2".
- SAE adaptors sizes #4 to #24.
- JIC adaptors sizes 1/4" to 1- 1/2".
- Code 61 adaptors sizes 1/2" to 1-1/2".
- Code 62 adaptors sizes 1/2" to 1-1/2".
- Optional Caterpillar®-style adaptors sizes 1/2" to 2" are available.

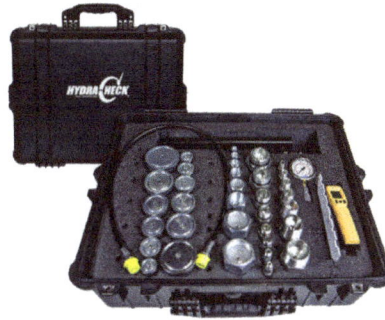

Fig. 2.3- Pressure/Leak Test Kit (Courtesy of Hydracheck)

4

4

2.4- Universal Flow Meter Test Kits

This kit covers a wide variety of flows from 1.0 GPM to 100 GPM (3.8 Lpm to 568 Lpm).

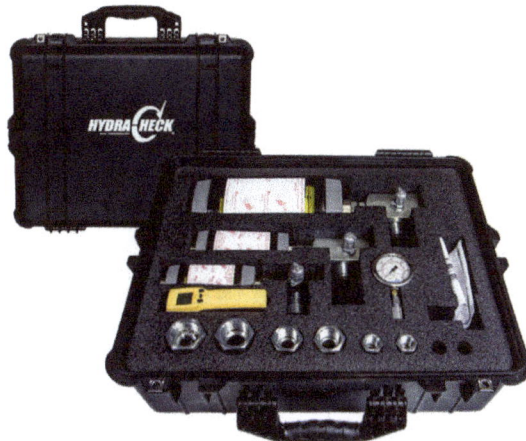

Fig. 2.4- Universal Flowmeter Test Kit (Courtesy of Hydracheck)

5

5

2.5- Portable Digital Hydraulic Multimeter

- All-in-one portable test unit.
- Test pumps, motors, valves and hydrostatic transmissions.
- Measure flow, pressure, peak pressure, temperature, power, and volumetric efficiency
- Bluetooth functionality.
- Onboard memory

Specifications:

- Pressures up to 7000 PSI (482 bar).
- Flows up to 210 GPM (950 lit/min).
- Flow Accuracy: ± 1%.
- Pressure Accuracy: ± 0.5% of full scale, Peak 1%.
- Temperature Accuracy: ± 2°F (± 1°C)
- Volumetric efficiency: ± 1%
- 1 ms response time.
- Choose between – Bar, PSI, MPa, & Ksc.

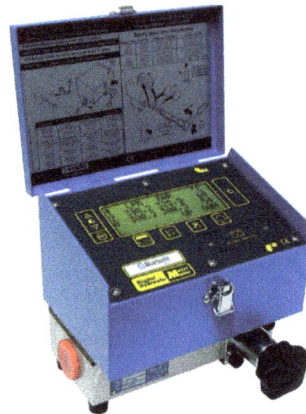

Fig. 2.5- Bidirectional Digital Hydraulic Multimeter with Bluetooth (Courtesy of Hydracheck)

6

6

2.6- Adaptor Kit

- The *kit* contains adapters needed to check pressure on most of hydraulic systems including Caterpillar, John Deere, Komatsu, Volvo, JCB, Genie, etc.

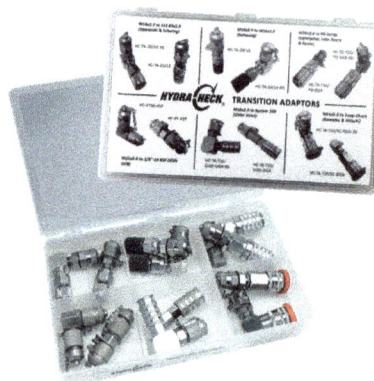

Fig. 2.6- Adaptor Kit (Courtesy of Hydracheck)

7

7

2.7- Test Points and Pressure Measurement Hoses

They are available in various sizes and configurations.

Fig. 2.7- Test Points and Measurement Hoses

8

8

2.8- Fluid Leakage Test Kit

- Detecting fluid leakage using *Florescent dyes*.

- Range of colors for various circuits.

- They work with any host fluid without damaging its properties.

- So, no need to drain the reservoir after detecting the leakage

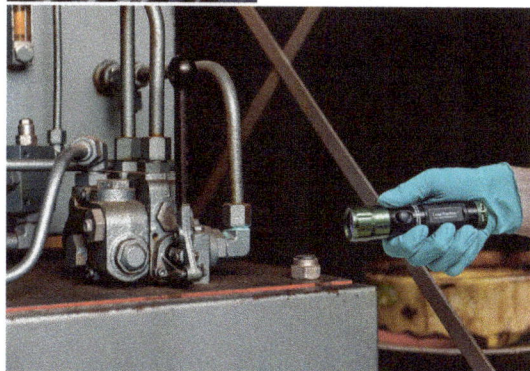

Fig. 2.8- Fluid Leakage Test Kit (Courtesy of Spectroline)

9

9

2.9- Surface Temperature Thermometers

- Touching hot surfaces is not a good practice, it can burn skin.
- Surface thermometers used to:
  - Detect internal leakage in hydraulic components.
  - Track faults that results in generating heat.

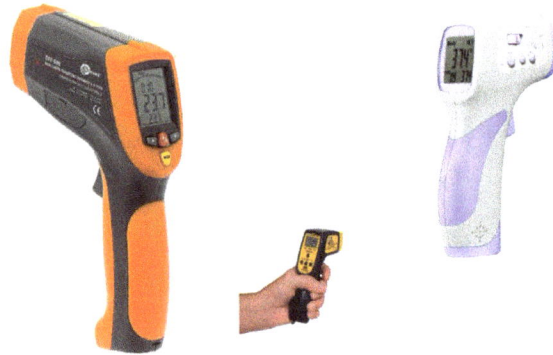

Fig. 2.9- Surface Temperature Thermometers

10

10

2.10- Vibration Indicators

Vibration is measured by:
- Permanently installed vibration sensor.
- Hand Arm Vibration Indicator (HVAI).

Fig. 2.10- Vibration Indicators

11

11

2.11- Tachometers

- RPM of rotating equipment (pumps or motors) is measure by:
- Permanently installed RPM sensor.
- mechanical or optical, and contact or contactless *tachometers*.

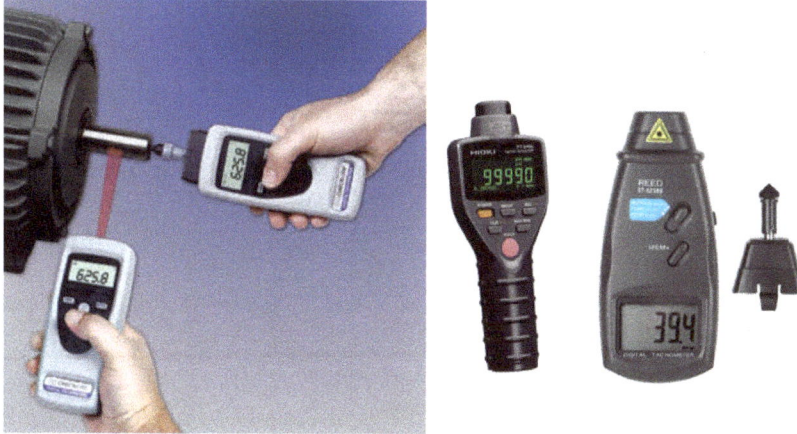

Fig. 2.11- Various Tachometers for Measuring RPM

12

12

2.12- Multimeters

Multimeter is used for checking Amperes, Volts, Resistances in electrohydraulic systems

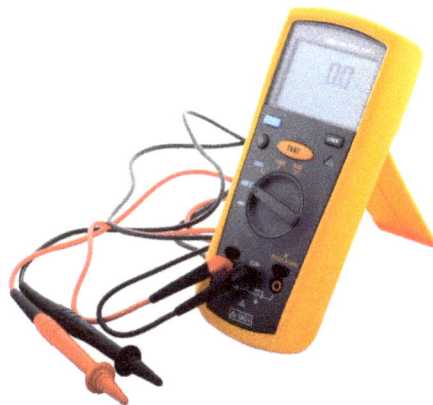

Fig. 2.12- Multimeter

13

13

2.13- Proportional Valve Tester

Suitable for:

- Control and functional testing of proportional valves with (OBE).
- Commissioning and troubleshooting of proportional valves easier.

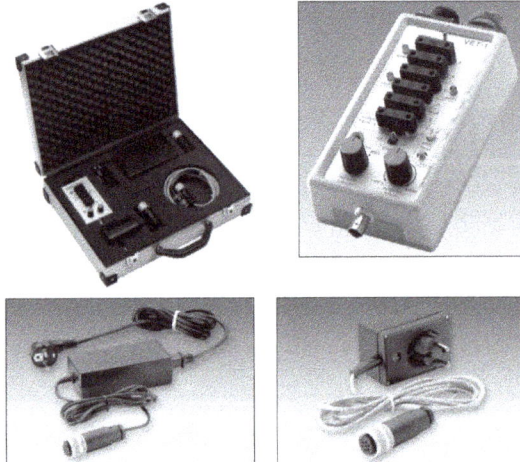

Fig. 2.13- Proportional Valve Tester Type VT-VETSY-1 (Courtesy of Bosch Rexroth)

14

14

Another example of *proportional valve tester* with the following features:

- Operation via touchscreen for digital or analog display.
- Automatic valve fault detection.
- Setpoint generator.
- Ramp function.
- Potentiometer operation.
- Two outputs for switching valves.

Caution:

- These testers should only be used by experienced persons who are familiar with the device, the valve, and the hydraulic system.

- Tester manufacturer assumes No liability will be accepted for damage caused by incorrect operation!

Fig. 2.14- Proportional Valve Tester Type VT-HDT-1-2X (Courtesy of Bosch Rexroth)

15

15

2.14- Servo Valve Tester

- Suitable for testing and commissioning of all proportional and servo proportional valves with **OBE**.

- Provides all command signals and measuring ports for time saving diagnosis of the valves.

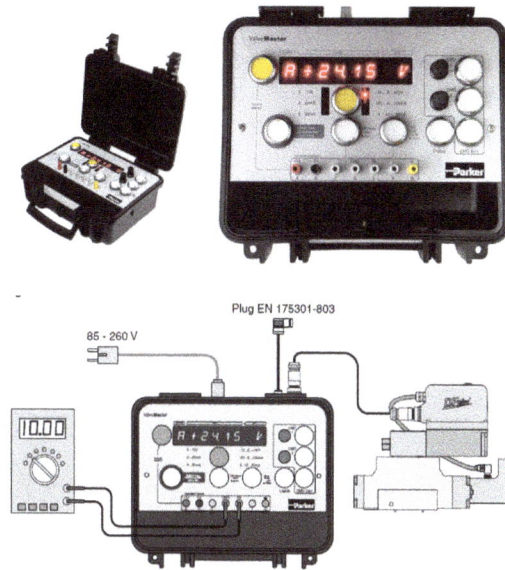

Plug EN 175301-803

85 - 260 V

**Fig. 2.15- Servo Valve Tester Series EX-M05
(Courtesy of Parker)**

16

16

Chapter 2 Reviews

1. Best practices to check on leaks, even small leaks, from various hydraulic circuits on one machine is?
   - A. Using hand under close to suspected transmission lines.
   - B. Run the machine at high speed and observe the leakage.
   - C. Run the machine and increase the load to a high pressure and observe the leakage.
   - D. Use the fluid leakage test kit with fluorescent dyes of different colors and light.

2. Which of the following instruments are used for checking the speed of the pump infield?
   - A. Multimeter.
   - B. Contactless Tachometer.
   - C. Surface Temperature Thermometer.
   - D. Vibration Indicator.

3. Which of the following instruments is used for checking the temperature of the pump infield?
   - A. Multimeter.
   - B. Contactless Tachometer.
   - C. Surface Temperature Thermometer.
   - D. Vibration Indicator.

4. Which of the following instruments is used for checking proper coupling of the pump infield?
   - A. Multimeter.
   - B. Contactless Tachometer.
   - C. Surface Temperature Thermometer.
   - D. Vibration Indicator.

5. Which of the following instruments is used for checking the electrical input signal to a solenoid operated valve infield?
   - A. Multimeter.
   - B. Contactless Tachometer.
   - C. Surface Temperature Thermometer.
   - D. Vibration Indicator.

Chapter 2 Assignment

Student Name: --- Student ID: ------------------

Date: -- Score: ------------------------

Question: suggest an equipment for checking and diagnoses the operation of a servo valve infield.

Chapter 3
Troubleshooting and Failure Analysis of Sealing Elements

Objectives:

This chapter presents guidelines for inspecting and troubleshooting hydraulic sealing elements. The chapter also presents 26 different failure modes, their causes and suggested solutions.

Brief Contents:

3.1- Hydraulic Seals Inspection

3.2- Hydraulic Seals Troubleshooting

3.3- Hydraulic Seals Failure Analysis

0

0

3.1- Hydraulic Seals Inspection

Hydraulic *sealing elements* could be:
- As simple as an O-Ring or a complex design of a piston seal package.
- Static or dynamic seals.
- Translational or rotational.

Volume 4 of this series of textbooks provides guidelines for sealing elements design, maintenance and safety.

1

1

Hydraulic Seals Inspection Sheet		Table 3.1
Manufacturer		
Model #		
Serial #		
Location		
Type of Seal	☐ Piston Seal ☐ Rod Seal ☐ Rotating Shaft Seal ☐ Other Seal	
Seal Failure	☐ Material of a hydraulic seal is harshly scorched. ☐ Seal is severely compressed and deformed in short time. ☐ Seal is extruded. ☐ Seal has abrasion market. ☐ Seal is leaking. ☐ Seal has short cuts. ☐ Seal is crushed and stressed beyond its limits and fails. ☐ Seal is squeezed and failed. ☐ Seal is hardened, became brittle, and has cracks. ☐ Seal is hardened, glazed, and has cracks. ☐ Seal has axial cuts particles embedded in the seal material. ☐ Seal lost its flexibility and cracks are formed. ☐ Seal softened, swell, or shrink. ☐ Seal material break-down, loss of physical properties, cracking, and crumbling. ☐ Seal has signs of bubble decompression. ☐ Seal has signs of Dieseling. ☐ Seal has increased gland clearance on one side only or uneven friction. ☐ Seal has excessive wear. ☐ Seal has torsional or spiral failure. ☐ V-Seal shows long cracks or splits.	2

2

3.2- Hydraulic Seals Troubleshooting

T-Seal-01-Seal Troubleshooting	
Material of a hydraulic seal is harshly scorched.	▪ Improper molding process.
Seal is severely compressed and deformed in short time.	▪ Poor seal material properties
Seal is extruded.	▪ Excessive Sealing Gap. ▪ Overpressure. ▪ Seal material is too soft. ▪ Improper seal size.
Seal has abrasion marks.	▪ Improper surface finish.
Seal is leaking.	▪ Improper seal design. ▪ Seal deterioration. ▪ Side loads.
Seal has short cuts.	▪ Seal pass over sharp edges during installation.

Table 3.2

3

Seal is crushed and stressed beyond its limits and fails.	▪ Seal is over pressurized.
Seal is squeezed and failed.	▪ Improper seal design (pressure trapping).
Seal is hardened, became brittle, and has cracks. Material splits through the seal body.	▪ Seal is overheated. ▪ Normal aging effect.
Seal is hardened, glazed, and has cracks.	▪ Seal is used at high speed.
Seal has axial cuts particles embedded in the seal material.	▪ High abrasive contamination.
Seal lost its flexibility and cracks are formed.	▪ Fluid incompatibility.
Seal softened, swell, or shrink.	▪ Chemical attach.
Seal material break-down, loss of physical properties, cracking, and crumbling.	▪ Exposure to water or emulsions.

4

Seal has signs of bubble decompression.	▪ Fluid aeration. ▪ Sudden decompression of air bubbles.
Seal has signs of Dieseling.	▪ Cavitation. ▪ Sudden decompression of air bubbles.
Seal has increased gland clearance on one side only or uneven friction.	▪ Side loading
Seal has excessive wear.	▪ Rough surface finish of gland. ▪ High working temperature. ▪ Poor fluid lubricity. ▪ Fluid incompatibility ▪ Mechanical vibration. ▪ Abrasive contamination. ▪ Sharp peaks and hard sealing surfaces.

5

Seal has torsional or spiral failure.	▪ Rough surface finish of gland. ▪ High working temperature. ▪ Poor fluid lubricity. ▪ High stroke speed. ▪ Long stroke. ▪ Side loads. ▪ Squeezed or soft seal. ▪ Too much space for movement in the groove. ▪ Type of metal surface. ▪ ID/W ratio of O-ring. ▪ Contamination or gummy deposits on metal surface. ▪ Eccentricity of sealing ring. ▪ Stretch of sealing rings. ▪ No use of Back-up Rings.
V-Seal shows long cracks or splits.	▪ Seal fatigue due to cold startup and/or exposure to cyclic pressure with high frequency

Video 369 (2.5 min)

6

6

3.3- Hydraulic Seals Failure Analysis

Challenges of Seals Failure:

▪ **Possible Remedies:** Seal damage is irreversible and there are no remedies.

▪ **Consequences of Seal Failure:**
   o A simple internal or external leakage.
   o Serious damage that create hazard for personnel and/or the public.
   o Critical applications: aircraft, amusement park rides, and elevating devices.

▪ **Cost of the seal vs. the Consequences:**

▪ **Failure Analysis:**
   o Visual inspection isn't enough.
   o Laboratory analysis is required.

A. Manufacturing Defects.
B. Seal and Gland (Groove) Design Issues.
C. Assembly Procedures.
D. Operational Conditions.
E. Normal Aging.
F. Storage Conditions.

7

7

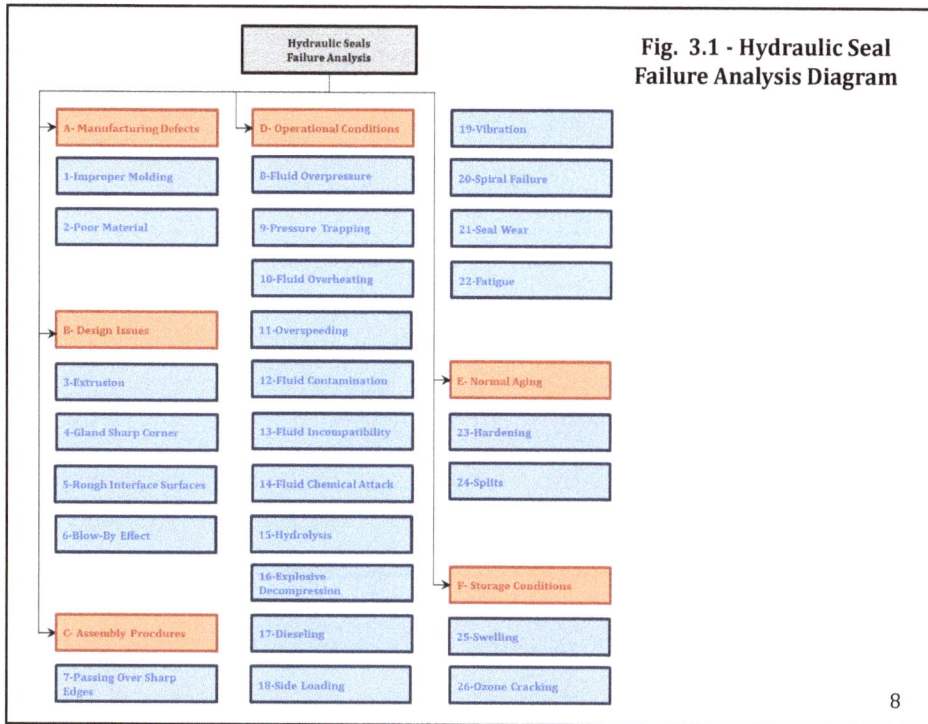

Fig. 3.1 - Hydraulic Seal Failure Analysis Diagram

8

3.3.1- Manufacturing Issues - Improper Molding

Failure Source: Improper *Molding* process due to defective dies, improper injection flow, pressure or temperature.

Failure Mode: Seal is harshly scorched.

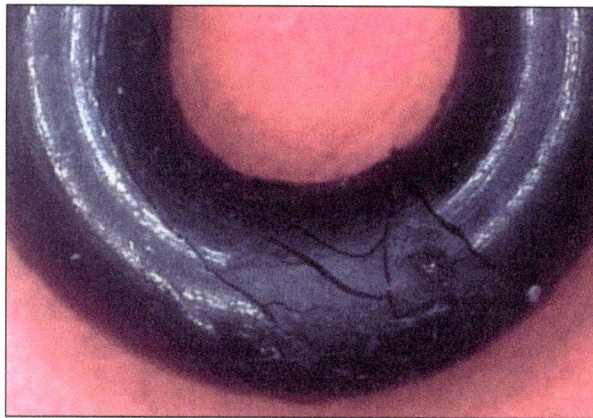

Fig. 3.2 - O-Ring Failure due to Improper Molding (www.o-ring-lab.com) 9

3.3.2- Manufacturing Defects - Insufficient Material Properties

Failure Source: Severe compression set due to poor material properties.

Failure Mode: a hydraulic seal is permanently deformed in short time.

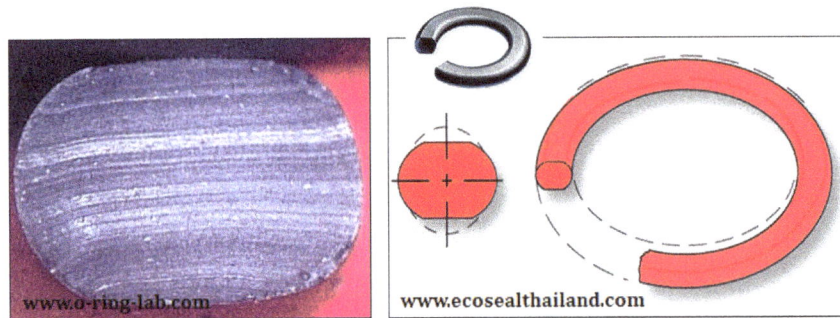

Fig. 3.3 - O-Ring Failure due to Insufficient Material Properties

10

10

3.3.3- Design Defects - Extrusion

Failure Source: extrusion and nibbling are caused by:

- **Excessive Sealing Gap:** Improper installation excessive clearances.
- **Nonuniform Sealing Gap:** Eccentricity → nonuniform clearance.
- **Less Hardness:** Seal material is too soft.
- **Degradation:** swelling, softening, shrinking, cracking, etc.
- **Improper Size:** Too large seal → excessive filling of groove.
- **Other Reasons:** overheating, side loading, soft material, and chemical incompatibility.

Extrusion Limits **are defined based on:**

- Seal hardness "Shore A",
- Sealing Gap "Diametral Clearance".
- Working pressure.

11

11

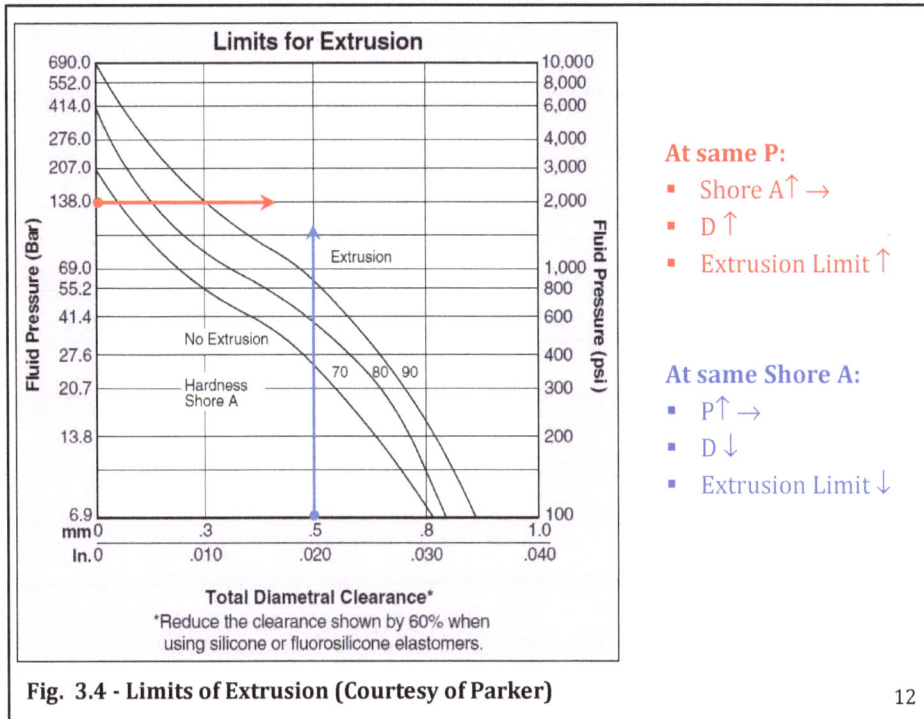

Limits for Extrusion

At same P:
- Shore A↑ →
- D ↑
- Extrusion Limit ↑

At same Shore A:
- P↑ →
- D ↓
- Extrusion Limit ↓

*Reduce the clearance shown by 60% when using silicone or fluorosilicone elastomers.

Fig. 3.4 - Limits of Extrusion (Courtesy of Parker)   12

12

Failure Mode: the O-Ring will extrude into the sealing gap.

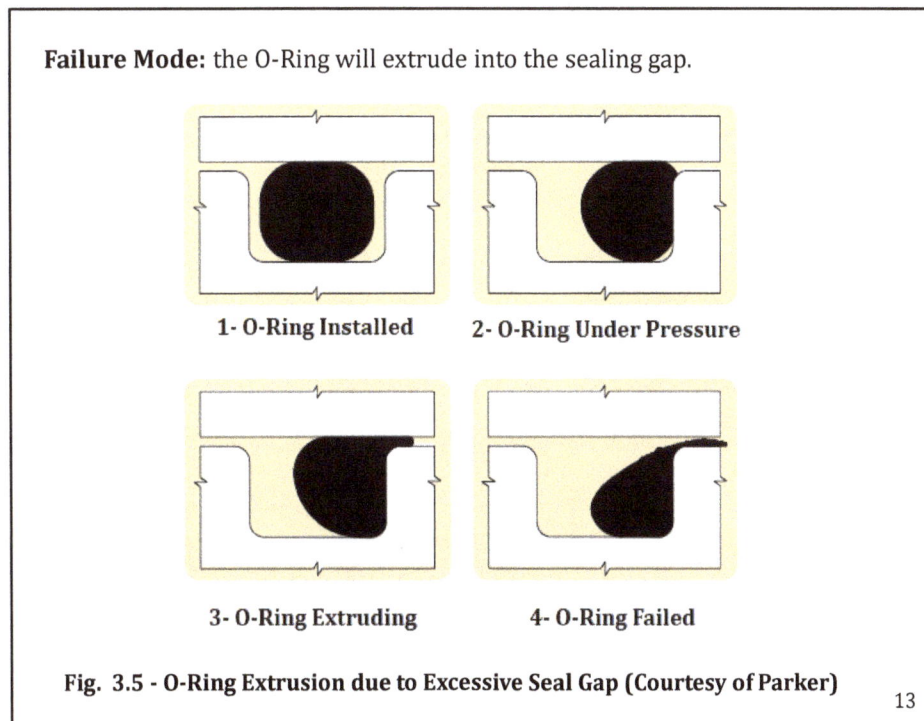

1- O-Ring Installed   2- O-Ring Under Pressure

3- O-Ring Extruding   4- O-Ring Failed

Fig. 3.5 - O-Ring Extrusion due to Excessive Seal Gap (Courtesy of Parker)   13

13

hydraulic seals creeping into the sealing gap. As a result, the seal deforms and/or breaks off.

Fig. 3.6 - Seal Extrusion, Example 1 (Courtesy of Parker)

14

14

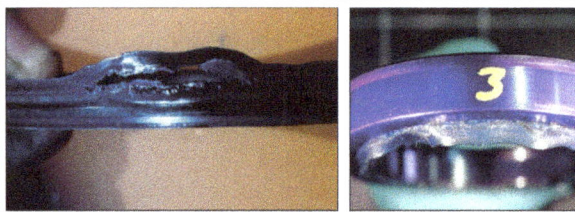

Fig. 3.7 - Seal Extrusion, Example 2

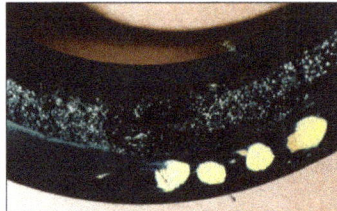

Fig. 3.8 - Seal Extrusion, Example 3 (Courtesy of System Seals Inc.)

15

15

Standard Test Method –
Extrusion-Resistance Test (ASTM C1183 / C1183M):
this method is used to determine the ability of a hydraulic seal to resist extrusion under certain pressure.

Suggested Solution:
- Solution 1-Proper Design of Sealing Gap.
- Solution 2-Use of Back-up Rings.

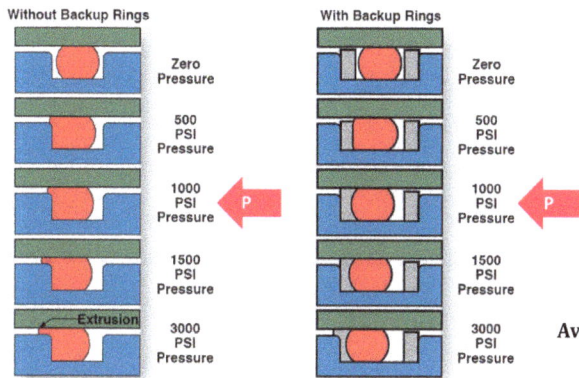

Fig. 3.9 - Seal Extrusion Avoidance by use of Back-up Rings (www.ecosealthailand.com)

16

16

- **Solution 3-Use of Anti-Extrusion Wedge-Rings:** It adjusts the gap dynamically as the pressure changes.

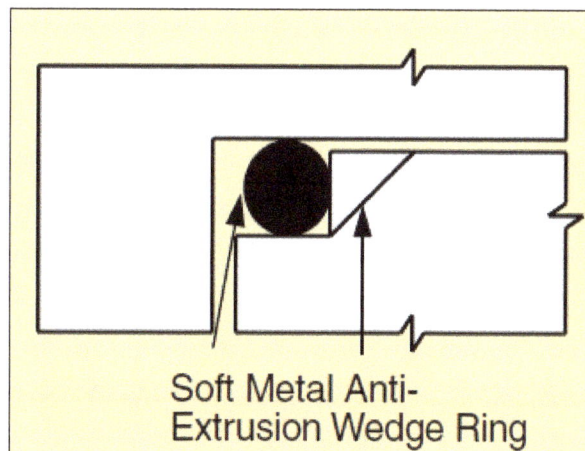

Soft Metal Anti-Extrusion Wedge Ring

Fig. 3.10 - Seal Extrusion Avoidance by use of Wedge-Rings (Courtesy of Parker)

17

17

3.3.4- Design Issues - Gland (Groove) Sharp Corners

Failure Source: Improper groove design with sharp corners.

Failure Mode: A seal groove isn't adequately rounded to the design code.
→ An O-Ring was at 250 bar (3,625 psi) was damaged at the outer circumference (even with relatively small seal gaps).

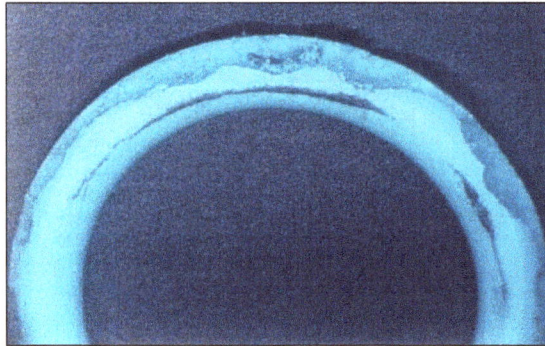

Fig. 3.11 - O-Ring Failure due to Sharp Corners and High Pressure
(www.o-ring-lab.com)

18

18

3.3.5- Design Issues - Rough Surfaces

Failure Source:
- Surface finish is important for dynamic seals.
- Rough surface → high friction with the seal material.
- Smooth surface → running the seal dry without lubrication.

Suggested Solution:
- Desirable surface finish is 10 - 20 micro-inches.
- Surface must be rough enough to hold small amounts of oil for lubrication.

- Surface finish < 5 micro-inches are not recommended for dynamic seals
  → an extending cylinder rod will be wiped completely dry.
  → and will not be lubricated when it retracts.

19

19

Failure Mode: Examples of seal abrasion due to contact with rough surface

Fig. 3.12

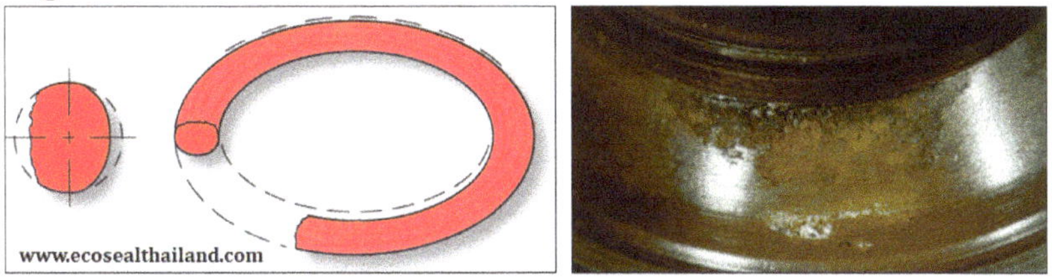

www.ecosealthailand.com

Fig. 3.13    Fig. 3.14 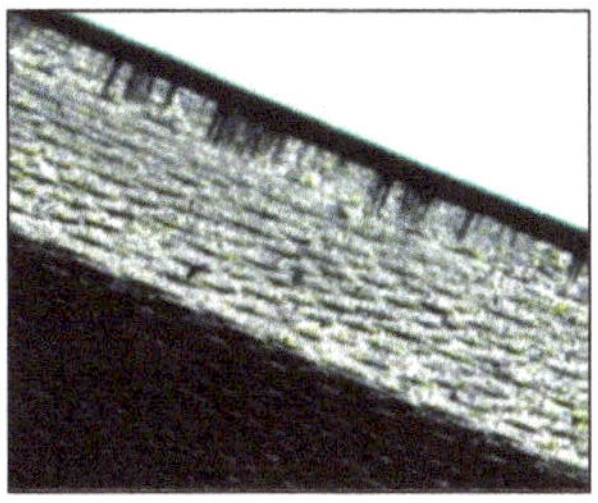   Fig. 3.15

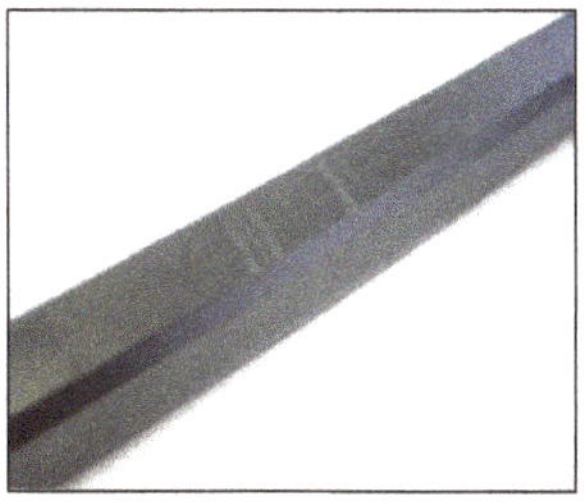

20

20

3.3.6- Design Issues - Blow-By Effect

Failure Source:
Traditional design of a bidirectional piston sealing solution
→ *Blow-By Effect* occurs → leakage.

Failure Mode: Increased leakage rate.

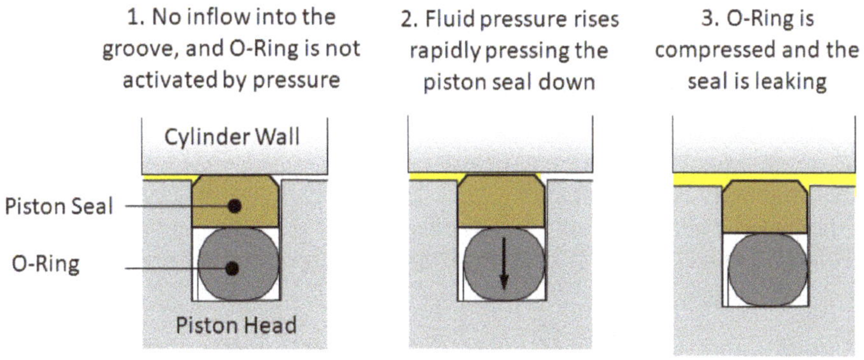

1. No inflow into the groove, and O-Ring is not activated by pressure

2. Fluid pressure rises rapidly pressing the piston seal down

3. O-Ring is compressed and the seal is leaking

Cylinder Wall

Piston Seal

O-Ring

Piston Head

Fig. 3.16 - Blow-By Effect

21

21

Suggested Solution:
Better design of piston seal with radial grooves or notches →
This controls dynamically the clearance → leakage controlled.

1. Fluid flows through radial groove

2. O-Ring compressed with the pressure increases
So that the leakage clearance is dynamically adjusted

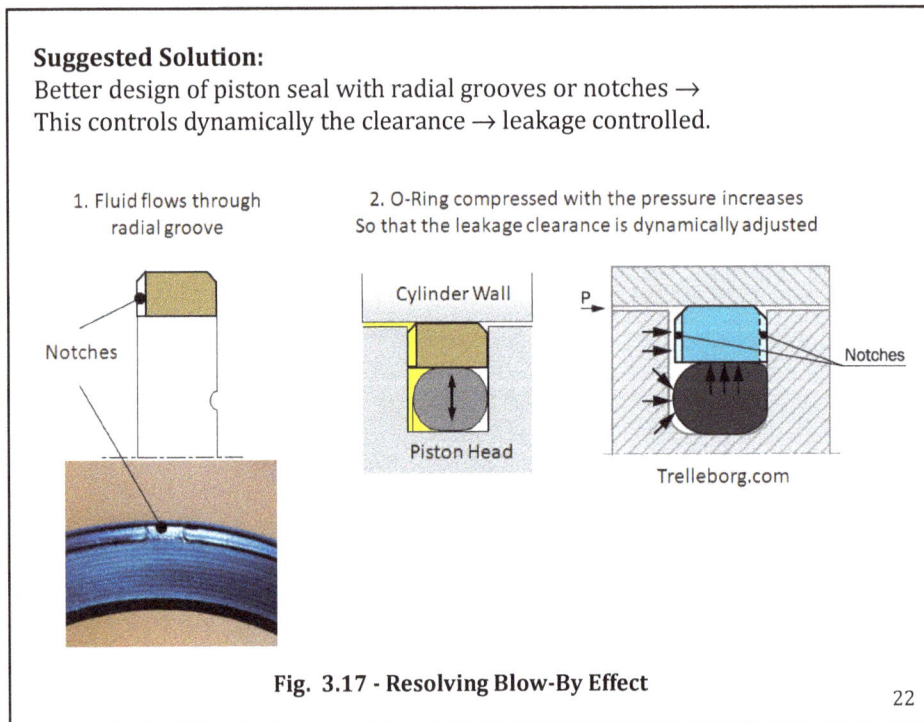

Fig. 3.17 - Resolving Blow-By Effect

22

22

3.3.7- Assembly Procedures - Passing Over Sharp Edges

Failure Source: Damage occurs during installation when:
- Seals come in contact with sharp edges such as threads.
- Insufficient lead-in chamfer.
- Oversize piston seal.
- Undersize rod seal.
- Seal is twisted/pinched during installation.
- Seal is not properly lubricated before installation.
- Seal is dirty or contaminated with metal particles upon installation.

Suggested Solution:
- Cover threads and sharp edges before assembly.
- Use of proper installation tools.
- Make sure lead in chamfers are based on manufacturers recommendations.
- Select proper seal size.
- Consider proper cleanliness during assembly.

23

23

Failure Mode:

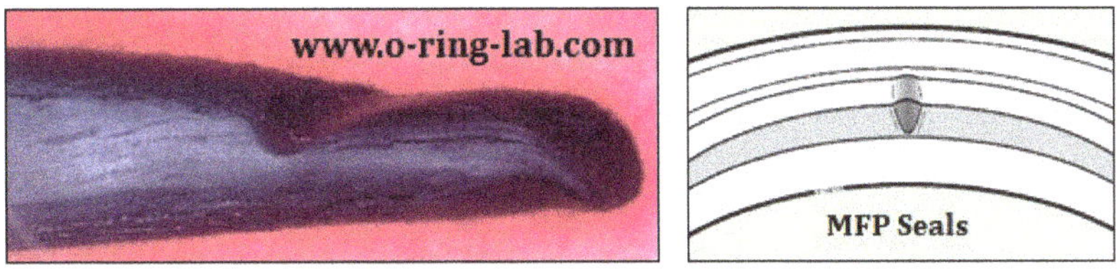

Fig. 3.18 - Sheared O-Ring due to Passing Over Sharp Edges in the
Gland Area During Assembly (www.o-ring-lab.com)

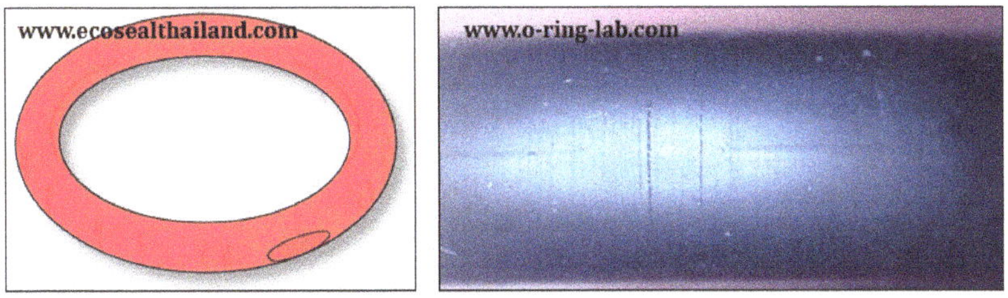

Fig. 3.19 - O-Ring Failure due to Passing Over Guide Chamfer

24

24

3.3.8- Operational Conditions - Overpressure

Failure Source: Overpressure.

Failure Mode: Hydraulic seals are stressed beyond their limits and fail.

Fig. 3.20

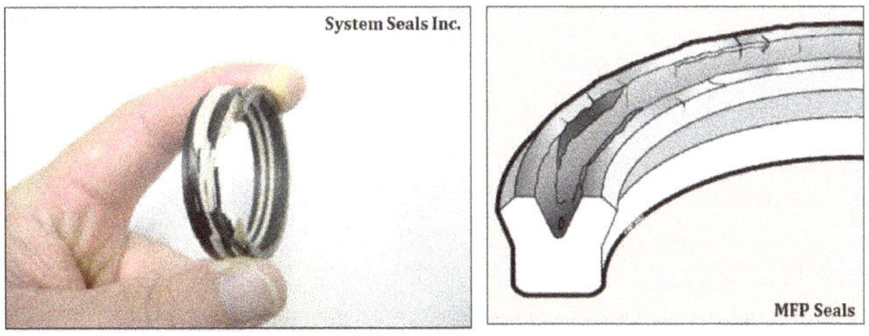

Fig. 3.21

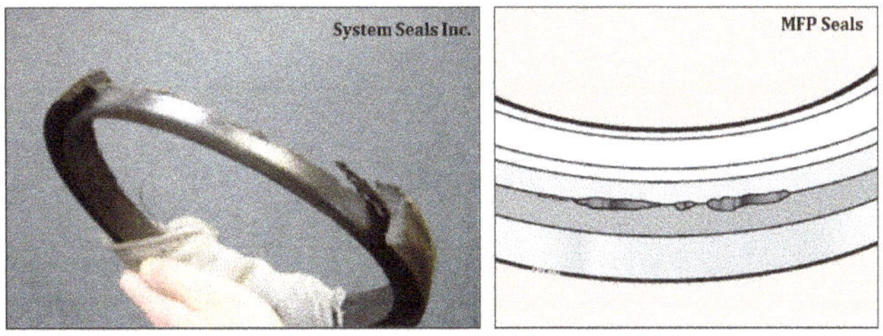

25

25

3.3.9- Operational Conditions - Pressure Trapping

Failure Source:

Pressure spikes → pressure is trapped between seals → seal damage.

Failure Mode: hydraulic seal is squeezed and failed

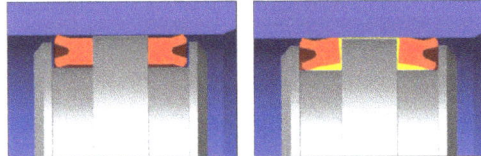

**Fig. 3.22 - Bidirectional Seal Failure due to Overpressure
(Courtesy of System Seals Inc.)**

Fig. 3.23 - Seal Failure due Pressure Trapping (Courtesy of System Seals Inc.)   26

26

3.3.10- Operational Conditions - Overheating

Failure Source:
- Long exposure to high working temperature or excessive heat.
- High speed operation that increases the seal lip temperature.

Failure Mode:

Overheating → Seal failures
- Seal hardness (particularly the sealing lip and the sliding surface)↑
- Cracks & Crack propagation over time.
- Softening & deformation of the seal body,
- Extrusion, splits, ruptures, melting, and squeezing.

Suggested Solution:
- Work within recommended temperature range.
- Select proper seal material that works better at high temperature.

27

27

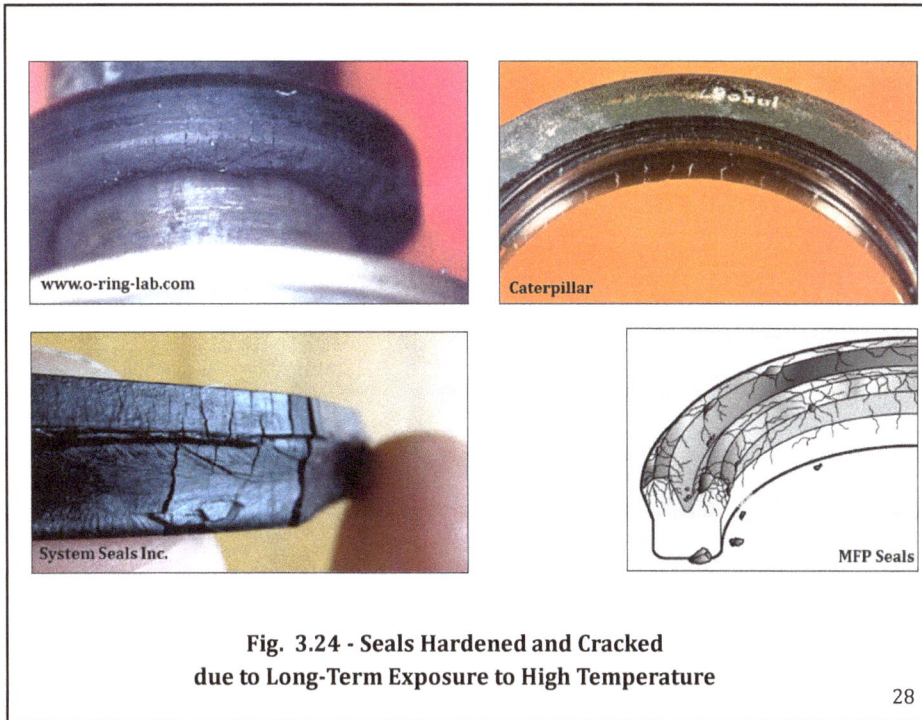

**Fig. 3.24 - Seals Hardened and Cracked
due to Long-Term Exposure to High Temperature**

28

28

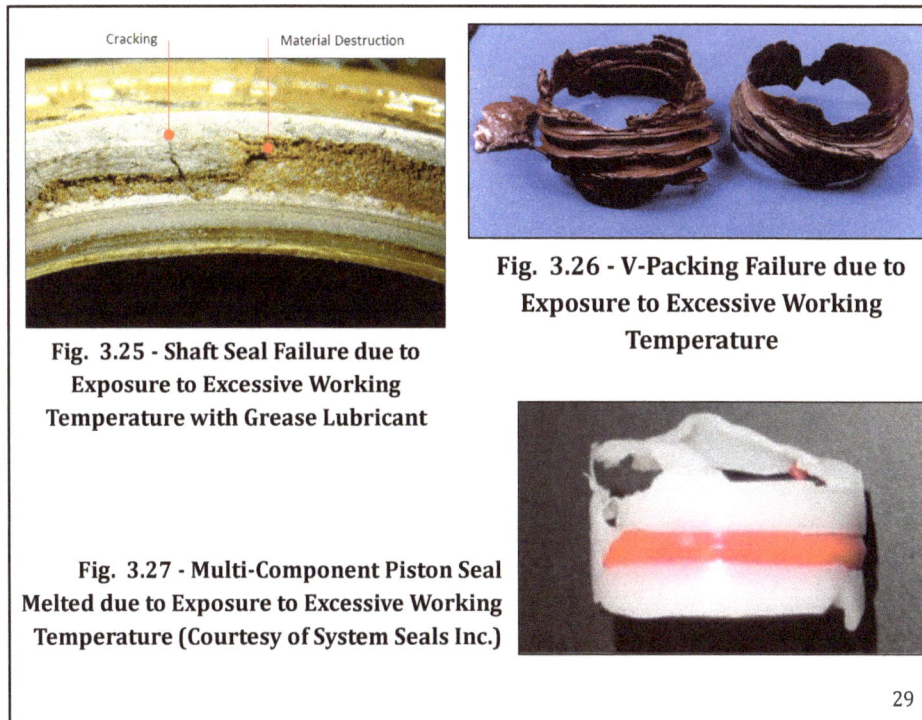

**Fig. 3.26 - V-Packing Failure due to
Exposure to Excessive Working
Temperature**

**Fig. 3.25 - Shaft Seal Failure due to
Exposure to Excessive Working
Temperature with Grease Lubricant**

**Fig. 3.27 - Multi-Component Piston Seal
Melted due to Exposure to Excessive Working
Temperature (Courtesy of System Seals Inc.)**

29

29

3.3.11- Operational Conditions – Over-Speeding

Failure Source: High surface speed generates excessive heat.

Failure Mode: Dynamic seal is hardened → cracks and glazing.

Suggested Solution:
- Reduce surface speed (stroke speed or RPM).
- Select proper seal material that works better at high speed and high temperature.

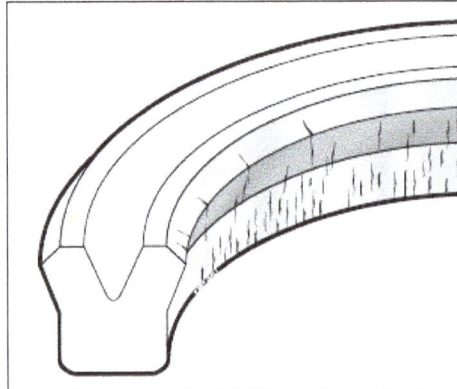

Fig. 3.28 - Hardened and Cracked Dynamic Sealing Surface due to Overspeeding (Courtesy of MFP Seals)

30

30

3.3.12- Operational Conditions - Contamination

Failure Source:
- Contaminated hydraulic fluid.
- Dirty assembly area.
- Poor wiper performance.
- Highly contaminated work environment.

Fig. 3.29 - Hydraulic Cylinder Operating in a Severe Salt Contaminated Work Environment (Courtesy of System Seals Inc.)

31

31

Failure Mode: Damaged seals due to contamination.

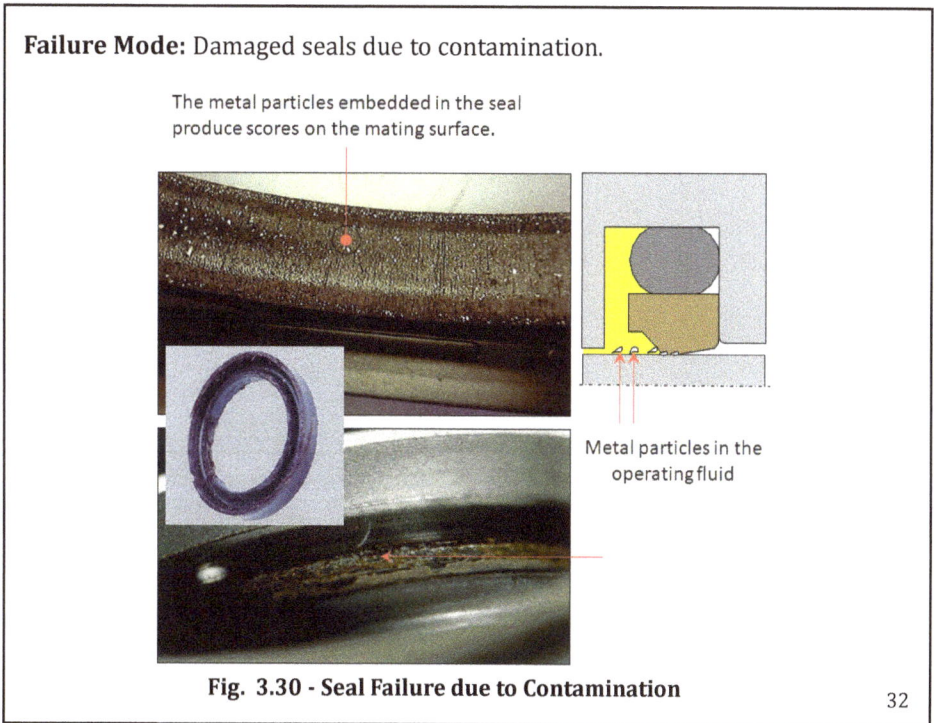

Fig. 3.30 - Seal Failure due to Contamination

32

32

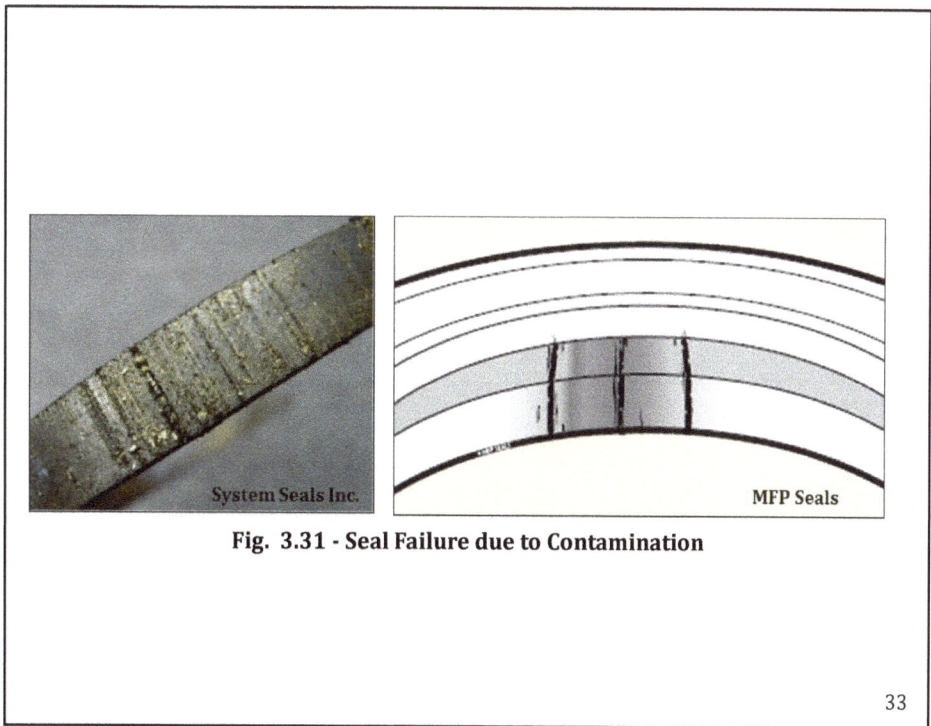

Fig. 3.31 - Seal Failure due to Contamination

33

33

3.3.13- Operational Conditions - Fluid Incompatibility

Failure Source: Incompatibility with the operating hydraulic fluid.

Failure Mode: Hydraulic seal lost its flexibility and cracks are formed.

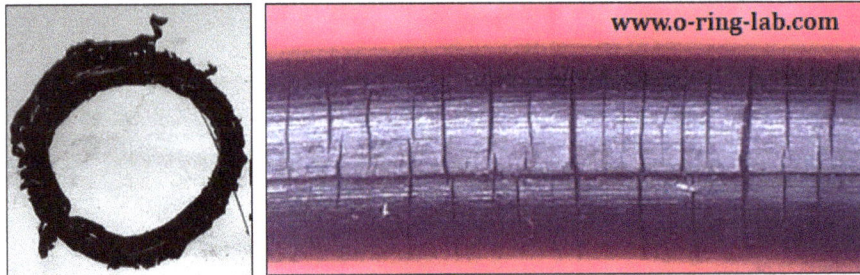

**Fig. 3.32 - Seal Failure due to Hydraulic Fluid Incompatibility
(www.o-ring-lab.com)**

34

3.3.14- Operational Conditions - Chemical Attack

Failure Source: Chemical interaction between the seal and the hydraulic fluid

Failure Mode: Excessive hardening, softening, swelling, and shrinkage.

Suggested Solution:
- Make sure acidity of the hydraulic fluid is within allowable limits.
- Check the resistivity level of the hydraulic seals to chemical attacks.

Fig. 3.33 - Seal Failure due to Hydraulic Fluid Chemical Attack

35

3.3.15- Operational Conditions - Hydrolysis

Failure Source: exposure to water or water-based fluids at elevated temperatures.

Failure Mode: Break-down of the seal material, loss of physical properties, cracking, and crumbling of the material due to *Hydrolysis* Failure.

Suggested Solution: Select proper seal material for water-based fluids.

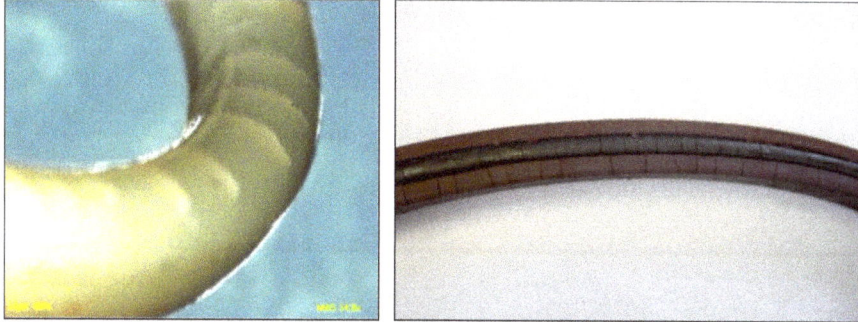

Fig. 3.34 - Seals Showing Early Signs of Hydrolysis
(Courtesy of System Seals Inc.)

36

36

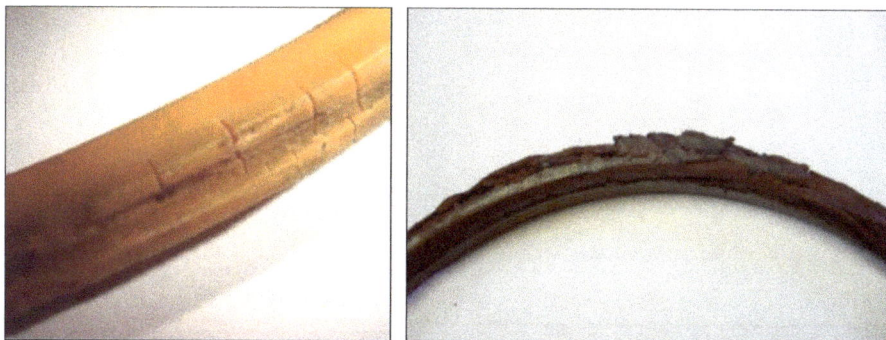

Fig. 3.35 - Seals Showing Late Signs of Hydrolysis
(Courtesy of System Seals Inc.)

37

37

3.3.16- Operational Conditions – Explosive Decompression

Failure Source:
Sudden pressure drop → Sudden decompression of air bubbles.

Standard Test Method (Explosive Decompression Test):
A high-pressure test rig is used to pressurize and depressurize the sealing element at a certain frequency under specified temperature.

Suggested Solution:

- Use *Anti-Explosive Decompression* (AED) seals.

- Apply the required design strategies to:

  o Prevent hydraulic fluid aeration and cavitation.

  o Control the decompression rate of pressurized fluid.

38

38

Failure Mode:

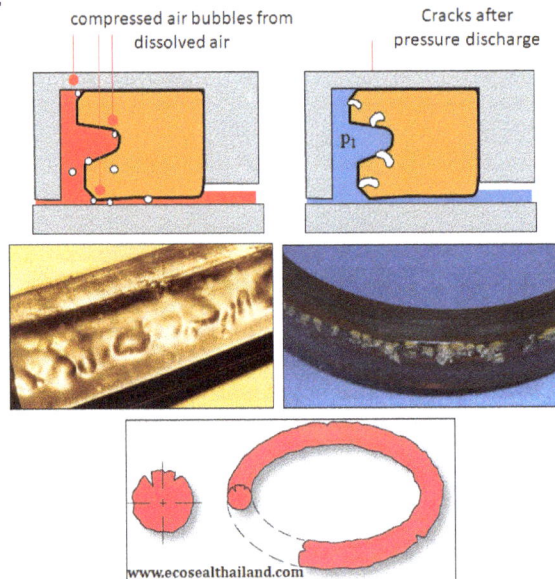

compressed air bubbles from dissolved air

Cracks after pressure discharge

P_1

www.ecosealthailand.com

Fig. 3.36 - Seal Failure due to Explosive Decompression, Example 1

39

39

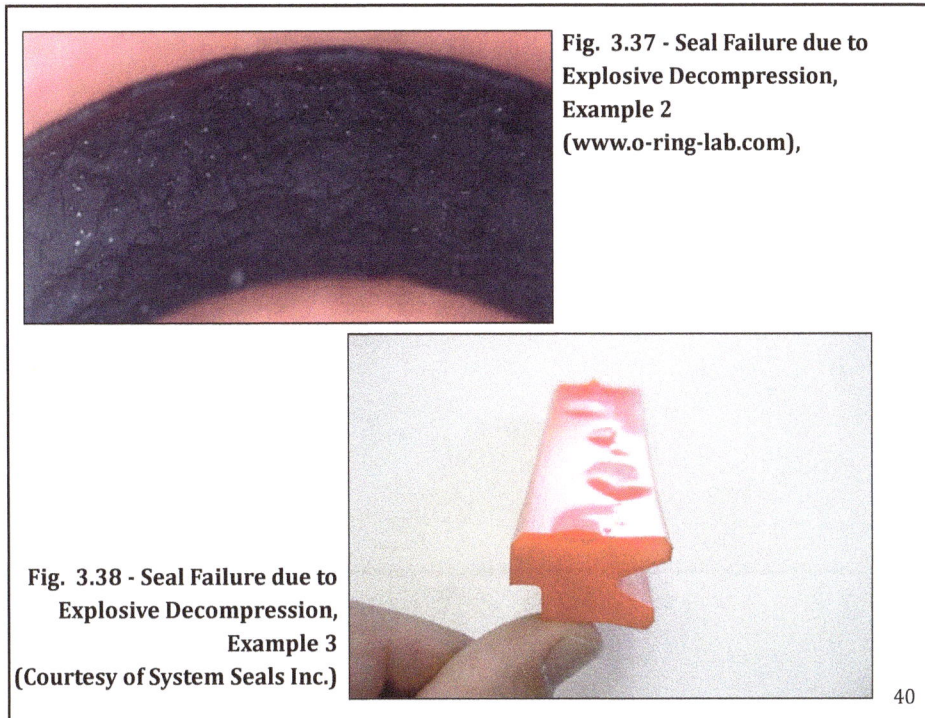

Fig. 3.37 - Seal Failure due to Explosive Decompression, Example 2 (www.o-ring-lab.com),

Fig. 3.38 - Seal Failure due to Explosive Decompression, Example 3 (Courtesy of System Seals Inc.)

40

40

3.3.17- Operational Conditions - Dieseling

Failure Source:
Aeration → air bubbles in the fluid→ high pressure
→ sudden **compression** of air bubbles → *Dieseling Effect*

Failure Mode: Seals are burned and damaged as a result of dieseling effect.

Suggested Solution: Apply the required design strategies:
  o Prevent hydraulic fluid aeration.
  o Control the working temperature and pressure shocks

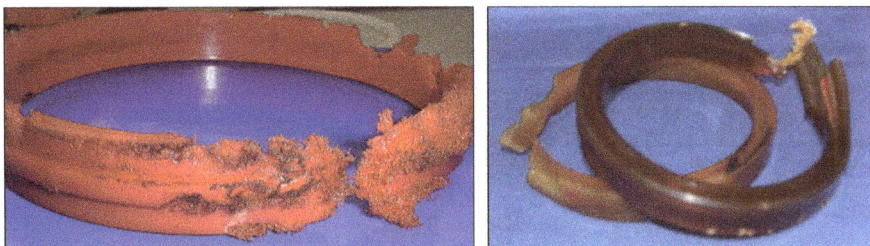

Fig. 3.39 - Seal Damage due to Diesel Effect, Example 1

41

41

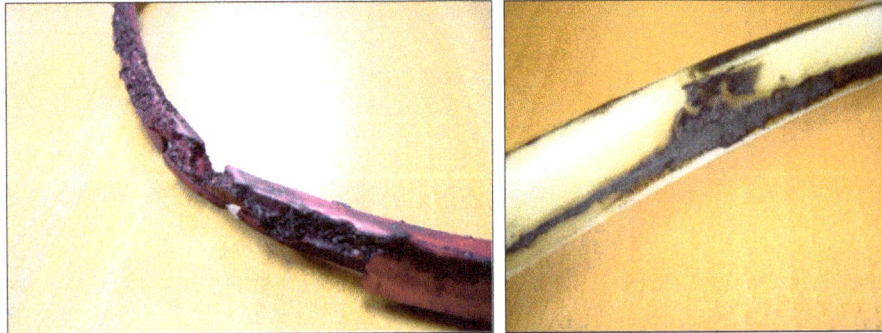

**Fig. 3.40 - Seal Damage due to Diesel Effect, Example 2
(Courtesy of System Seals Inc.)**

42

42

3.3.18- Operational Conditions - Side Loading

Failure Source:
- Side loads on a cylinder piston or cylinder rod.
- Insufficient load guidance.

Suggested Solution:
- Follow the best practices for mounting the hydraulic cylinder and the attached load.

- Whenever possible, use *Stop Tubes* to reduce the effect of side loads.

- Proper seal package design that includes Guide-Rings.

43

43

Failure Mode:

- Damage to the sealing elements and the cylinder components
- Increased gland clearance on one side only.
- Gap extrusion.
- Increased leakage.
- Uneven friction on the seal.
- Rod or barrel will be galled or scored.

Severe side loading with catastrophic damage from metal-to-metal contact

Side loading that wore the chrome off the rod

**Fig. 3.41 - Seal Damage due to Side Loading
(Courtesy of System Seals Inc.)**

44

44

3.3.19- Operational Conditions - Vibration

Failure Source: small frequent motions and vibration which are usually encountered when equipment is in transit. Such defects are reported in hydraulic cylinders more than any other components.

Failure Mode: Excessive wear of hydraulic seals.

Suggested Solution: Apply the required design strategies to isolate the vibrations from hydraulic components.

45

45

3.3.20- Operational Conditions - Spiral Failure

Failure Source: Spiral failure occurs when some segments of the O-ring slide while other segments simultaneously roll. Reasons are one or more of:

1. Speed of stroke.
2. Lack of lubrication.
4. Squeeze and softness of sealing rings.
5. too much space for movement in the groove.
6. Temperature of operation.
7. Length of stroke.
8. Surface finish of gland.
9. Type of metal surface.
10. Side loads.
11. ID to W ratio of O-ring.
12. Contamination or gummy deposits on metal surface.
15. Eccentricity of sealing ring.
16. Stretch and softness of sealing rings.
17. Lack of Back-up Rings.

46

46

Failure Mode:

- A unique type of failure (*Torsional or Spiral Failure*).
- Occur on reciprocating dynamic sealing rings.
- 45-degree angle deep cuts through the crosssection in a spiral pattern.

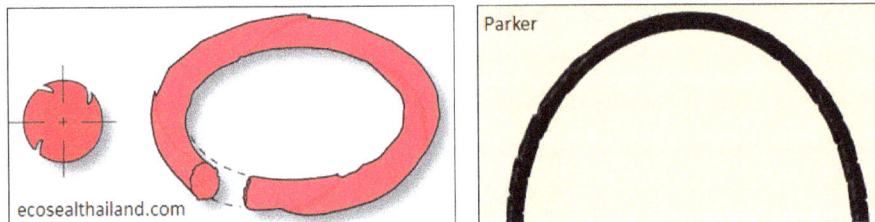

Fig. 3.42 - Sealing Ring Spiral Failure
(ecosealthailand.com)

47

47

3.3.21- Operational Conditions - Seal Wear

Failure Mode:

Factors that Influence Seal Wear	
Rough surface finish	Excessive abrasion may occur
Ultra-smooth surface finish	Surface finishes below 2 μm Ra can create aggressive seal wear due to lack of lubrication
High pressure	Increases the radial force of the seal against the dynamic surface
High temperature	While hot, materials soften, thus reducing tensile strength
Poor fluid lubricity	Increases friction and temperature at sealing contact point
Tensile strength of seal compound	Higher tensile strength increases the material's resistance to tearing and abrading
Fluid incompatibility	Softening of seal compound leads to reduced tensile strength
Coefficient of friction of seal compound	Higher coefficient materials generate higher frictional forces
Abrasive fluid or contamination	Creates grooves in the lip, scores the sealing surface and forms leak paths
Extremely hard sealing surface	Sharp peaks on hard surfaces will not be rounded off during normal contact with the wear rings and seals, accelerating wear conditions

Table 3.3 - Factors Affecting Seal Wear (Courtesy of Parker)

48

48

Failure Mode: (uneven, grooved, or excessive) wear in dynamic seals.

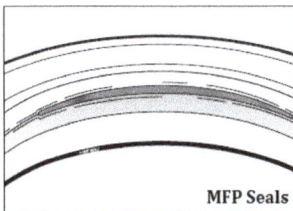

The dynamic lip is worn to a rounded, egg-shape.

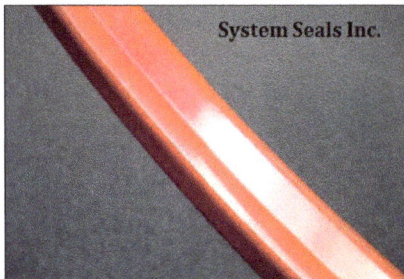

Only one side of the dynamic lip is showing excessive wear.

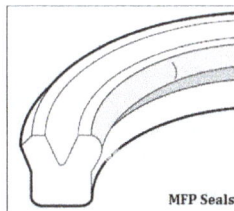

Polyurethane Rod Seal with Shiny and Smooth Surface from a Dry Running Condition

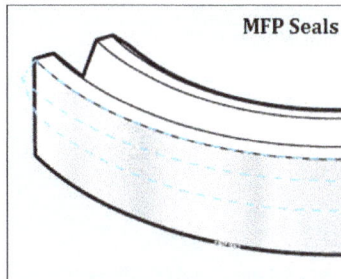

The dynamic face of the seal is worn to a glossy mirror like shine.

49

49

3.3.22- Operational Conditions - Fatigue

Failure Source: Cold startup and/or Exposure to cyclic pressure with high frequency.

Failure Mode: The V portion of the seal shows long cracks or splits.

Suggested Solution:
- Proper selection of seal design and material.
- Control the startup temperature.

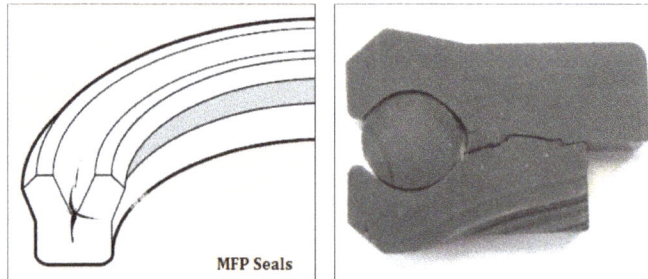

Fig. 3.44 - Profile Section of an O-Ring Loaded U-Cup with Flex Fatigue Cracking (Courtesy of System Seals Inc.)

50

50

3.3.23- Normal Aging - Hardening

Failure Source: Normal *Aging*.

Failure Mode: Cracks on the inner circumference of the sealing ring.

Suggested Solution:
- Use the seals within their estimated lifetime.
- Make sure storage is arranged based of first-come first-use basis.

Hardening cracks in the sealing lip

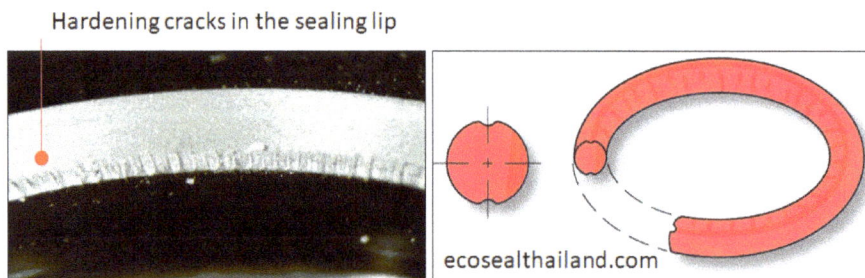

Fig. 3.45 - Sealing Ring Hardening due to Normal Aging

51

51

3.3.24- Normal Aging - Splits

Failure Source: Expired *Shelf Life*.

Failure Mode: Material split through the seal body.

Suggested Solution:
- Use the seals within their estimated lifetime.
- Make sure storage is arranged based of first-come first-use basis.

Fig. 3.46 - Failure due to Normal Aging (www.o-ring-lab.com)

52

52

3.3.25- Storage Conditions - Swelling

Failure Source: Seal swilling could occur due to:
- Elastomers have a higher coefficient of thermal expansion than steel.
- Fluid incompatibility or chemical attack.
- Absorbing moisture.

Failure Mode: When a seal absorb the surrounding water like a sponge, it swells to the point of malfunction.

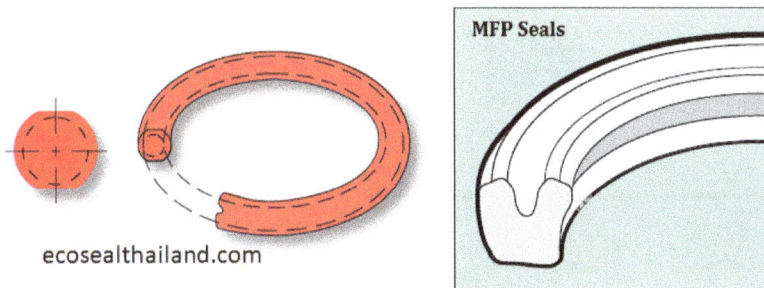

Fig. 3.47 - Sealing Ring Excessively Swells due to Humid Storage Space (ecosealthailand.com)

53

53

Suggested Solution:
- Adjust the humidity level at the storage space.
- Test the volume change of hydraulic seals periodically.

Fig. 3.48 - Swelling Failure of a wiper
(Courtesy of System Seals Inc.)

54

54

3.3.26- Storage Conditions - Ozone Cracking

Failure Source: This damage is a result of exposure of a sealing ring to Ozone for several weeks without protection.

Failure Mode: Many small surface cracks perpendicular to the direction of stress.

Suggested Solution: Provide proper protection against Ozone.

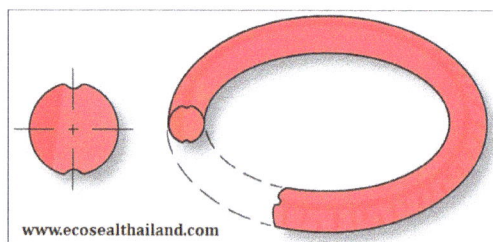

Fig. 3.49 - Sealing Ring Surface Cracking due to Exposure to Ozone
(ecosealthailand.com)

55

55

Chapter 3 Reviews

1. Which of the following causes seals extrusion?
   A. Excessive Sealing Gap.
   B. Overpressure.
   C. Seal material is too soft.
   D. All of the above mentioned.

2. Which of the following causes a seal to be hardened, glazed, and has cracks?
   A. The seal is used at high speed.
   B. Hydraulic fluid is incompatible with the seal.
   C. The seal is exposed to water.
   D. The seal is exposed to cavitation.

3. Which of the following causes signs of dieseling to show on a seal?
   A. The seal is used at high speed.
   B. Hydraulic fluid is incompatible with the seal.
   C. The seal is exposed to water.
   D. The seal is exposed to cavitation.

4. Which of the following causes axial cuts on a seal with embedded particles in the seal material?
   A. The seal is used at high speed.
   B. Hydraulic fluid is incompatible with the seal.
   C. The seal is exposed to overpressure.
   D. The fluid is contaminated by abrasive particulate contaminants.

5. The shown below seal failure is likely due to which of the following?
   A. Overpressure and excessive seal gap.
   B. Aeration of the oil.
   C. Abrasive contaminants.
   D. Lack of lubricatio.

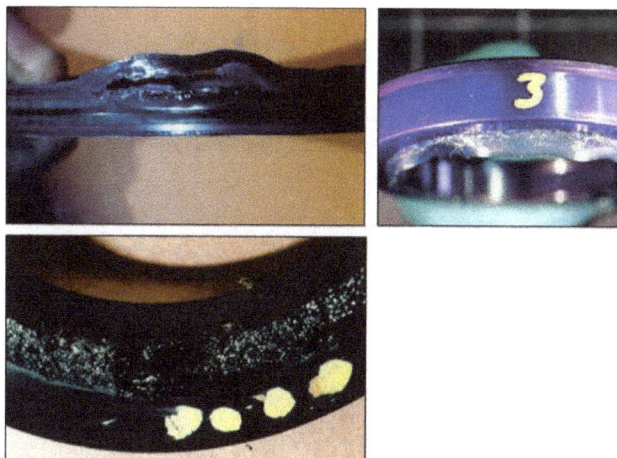

Chapter 3 Assignment

Student Name: --- Student ID: ------------------

Date: -- Score: ------------------------

Question: List reasons for spiral failure of a hydraulic seal

Chapter 4
Troubleshooting and Failure Analysis of Pumps

Objectives:

This chapter discusses hydraulic *pumps* inspection, troubleshooting, and failure analysis. In this chapter, troubleshooting charts for twelve different faults of hydraulic pumps are presented. The chapter also presents examples of defective pumps due to contamination, overheating, cavitation, and fatigue stress for gear, vane, and piston pumps.

Brief Contents:

4.1- Hydraulic Pumps Inspection

4.2- Hydraulic Pumps Troubleshooting

4.3- Hydraulic Pumps Failure Analysis

0

0

4.1- Hydraulic Pumps Inspection

Hydraulic *pumps* have various characteristics based on:

- **Displacement:** Fixed or Variable.
- **Rotation:** Unidirectional, Bidirectional, or Over Center.
- **Mechanism:** Gear, Vane, and Piston.

Volume 1 of this series of textbooks presents an overview about the construction and operating principle of various pump mechanisms.

1

1

Hydraulic Pump Inspection Sheet	
Manufacturer	
Model #	
Serial #	
Location	
Pumping Mechanism	☐ External Gear ☐ Internal Gear ☐ Gerotor ☐ Vane Pump [☐ Balanced ☐ Unbalanced] ☐ Radial Piston [☐ Rotating Cam ☐ Rotating Cylinder Block] ☐ Bent Axis ☐ Swash Plate ☐ Other [ ]
Direction of Rotation	☐ Unidirectional ☐ Bidirectional ☐ Over Center
Pump Displacement	☐ Fixed ☐ Variable [= cc/rev]
Type of Control	☐ Pressure Compensated ☐ Displacement Controlled ☐ Constant Power (Torque) ☐ Load Sense
Drive Shaft	Type and Size:
Ports	Case Drain: ☐ Yes ☐ NO Case Drain size: Inlet Port size: Outlet Port Size:
Conditions of Seals	
Conditions of Bearings	
Conditions of Inside Parts	
Other Nots	

Table 4.1 – Hydraulic Pumps Inspection Sheet

2

2

4.2- Hydraulic Pumps Troubleshooting
4.2.1- No Flow Out of the Pump

T-Pump-01-No Flow out of the Pump	
Pump drive motor is not working.	▪ Check wiring connection to electric motor. ▪ Check wiring connection in control circuit. ▪ Check use of correct voltage. ▪ Check fuses in the electrical control circuit.
Pump rotates in wrong direction.	▪ Check polarity of electrical motor.
Pump is not receiving fluid.	▪ Check oil level in the reservoir. ▪ Inspect intake line for restriction, kinking, or closed suction valve. ▪ Check clogged strainers or suction filters.

**Table 4.2- Troubleshooting Chart
(T-Pump-01-No Flow out of the Pump)**

3

3

Video 180 (0.5 min)

Is it a variable displacement pump?	▪ Check the setting of the pump controller.
Pump-motor coupling is sheared. Video 181 (0.5 min)	▪ Check pressure spikes. ▪ Check coupling misalignment.
Pump is not primed.	▪ Prime the pump after reviewing the relevant instructions.
Pumping elements are severely worn, damaged or seized?	▪ Stuck internal components from varnish in the oil or from rust and corrosion. Varnish indicates the system is running too hot. Rust or corrosion may indicate water in the oil.
Pump is wrongly assembled. **(See Note 1).**	▪ Assemble the pump according to the manufacturer assembly instructions.

Table 4.2- Continue

4

4

Note 1: An example of an incorrectly assembled pump

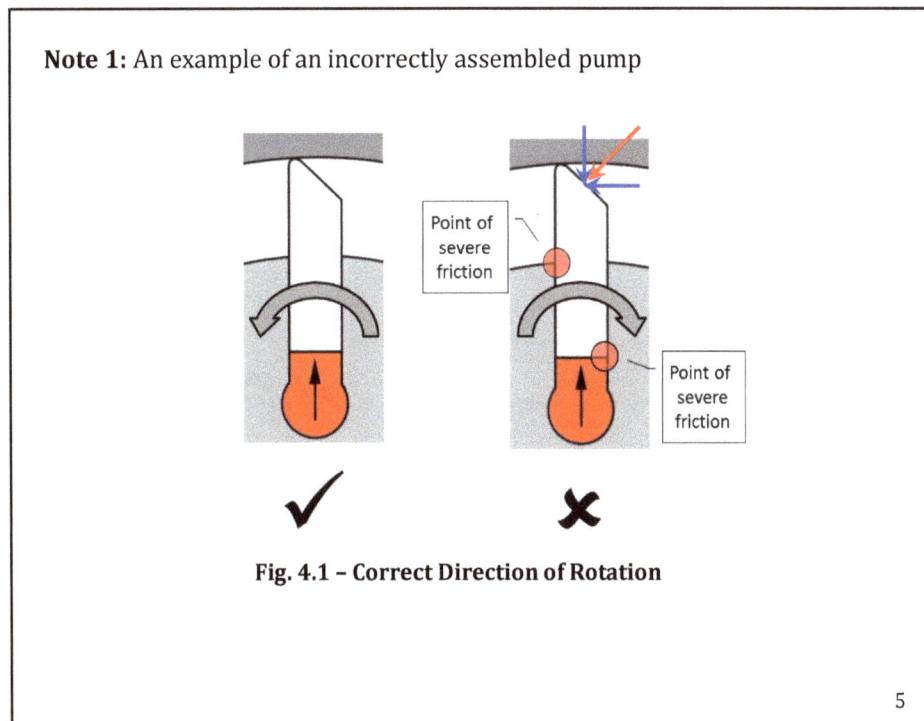

Fig. 4.1 – Correct Direction of Rotation

5

5

4.2.2- Low Flow Out of the Pump

Video 183 (0.5 min)

T-Pump-02-Low Flow out of the Pump	
Pump is recently changed.	▪ Review the pump ordering code.
Pump rotates at low speed.	▪ Check the drive motor speed.
Pump is driven by 3-phase electric motor.	▪ Check fuses on all three phases of a 3-phase electric motor. If the fuse on one phase is blown the motor may run but will overheat and will not produce full power.
Is it a variable displacement pump?	▪ Check the setting of the pump controller. ▪ Check over pressure for pressure compensated pumps.
Pump cavitation.	▪ Consult Chart: "**T-System-02-Pump Cavitation**".
Pump worn.	▪ Conduct pump performance test **(see Note 1)** to check internal leakage rate and then rebuild or replace the pump accordingly.

Table 4.3- Troubleshooting Chart
(T-Pump-02-Low Flow out of the Pump)

6

6

Note 1:
- Pump A and pump B both show the ability to achieve certain pressure.
- Pump B is very inefficient.

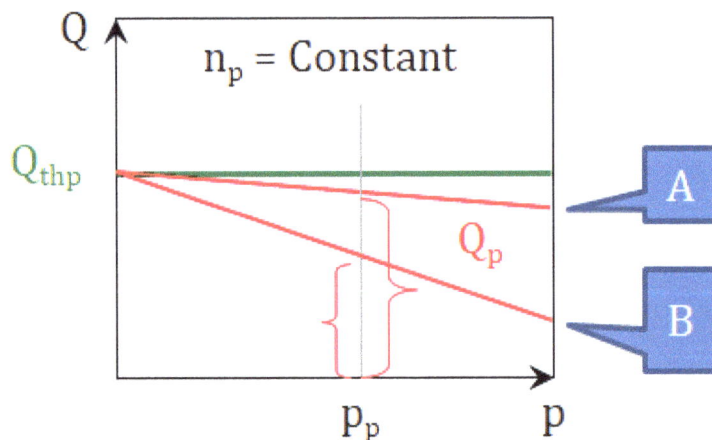

Fig. 4.2 – Pump Test Results

7

4.2.3- Erratic Flow Out of the Pump

T-Pump-03-Erratic Flow out of the Pump	
Fluid Aeration.	▪ Consult Chart: ▪ "T-System-01-Fluid Aeration".
Pump Cavitation.	▪ Consult Chart: ▪ "T-System-02-Pump Cavitation".
Is it a variable displacement pump?	▪ Check the setting of the pump controller.
Worn or inefficient pump.	▪ Test the pump.
Excessive pump wear.	▪ Consult Chart: ▪ "T-Pump-10-Excessive Pump Wear".

**Table 4.4- Troubleshooting Chart
(T-Pump-03-Erratic Flow out of the Pump)**

8

8

4.2.4- Excessive Flow Out of the Pump

T-Pump-04-Excessive Flow out of the Pump	
Pump is oversized.	▪ Review the pump ordering code. ▪ Check if cam ring of a vane pump is worn.
Pump rotates at high speed	▪ Check the drive motor speed.
Is it a variable displacement pump?	▪ Check the setting of the pump controller.

**Table 4.5- Troubleshooting Chart
(T-Pump-04-Excessive Flow out of the Pump)**

9

9

4.2.5- No Pressure at the Pump Outlet

T-Pump-05-No Pressure at the Pump Outlet	
No flow out of the pump?	▪ Consult Chart: ▪ **"T-Pump-01-No Flow out of the Pump"**.
Pressure gauge is faulty?	▪ Replace the pressure gauge.
Pressure relief valve is left fully opened.	▪ Reset the valve.
Is it a variable displacement pump?	▪ Check the setting of the pump controller.
Circuit design issues	▪ Check if the pump is vented through an open or tandem center valve? ▪ Check pump unloading method.

Table 4.6- Troubleshooting Chart
(T-Pump-05-No Pressure at the Pump Outlet)

10

10

4.2.6- Low Pressure at the Pump Outlet

Video 193 (0.5 min)

T-Pump-06-Low Pressure at the Pump Outlet	
Pressure relief valve is set too low.	▪ Reset the valve.
Is it a variable displacement pump?	▪ Check the setting of the pump controller.
External load is below normal or no load?	▪ Check load conditions.
Leaking Pump?	▪ Consult Chart: ▪ **"T-Pump-09-Leaking Pump"**.
Pressure relief valve is at fault?	▪ Consult Chart: ▪ **"T-Valve-03-PCV Troubleshooting"**.

Table 4.7- Troubleshooting Chart
(T-Pump-06-Low Pressure at the Pump Outlet)

11

4.2.7- Erratic Pressure at the Pump Outlet

T-Pump-07-Erratic Pressure at the Pump Outlet	
Erratic flow out of the pump?	■ Consult Chart: ■ "T-Pump-03-Erratic Flow out of the Pump".
Excessive pump wear?	■ Consult Chart: ■ "T-Pump-10-Excessive Pump Wear".
Pressure relief valve is at fault (worn or sticking relief valve)?	■ Consult Chart: ■ "T-Valve-03-PCV Troubleshooting".
A pump works with an accumulator in parallel?	■ Check the accumulator for gas leak or damage of accumulator piston, bladders, or diaphragm.
Backup pressure relief valve set near setting of a variable pump controller.	■ Set relief valve at 10% higher than the variable pump compensator (See Note 1).
Shuddering of overrunning load controlled by PO Check?	■ Check sizing of the PO Check ■ Valve and review circuit design (See Note 2).

Table 4.8- Troubleshooting Chart
(T-Pump-07-Erratic Pressure at the Pump Outlet)

12

12

Note 1:
- backup relief valve set 10 bar (150 psi) > pump compensator.
- Otherwise → pump controller frequently actuate → erratic pressure.

Note 2: pump, cylinder, and the PO check valve aren't compromisingly sized → load shuddering → erratic pressure.

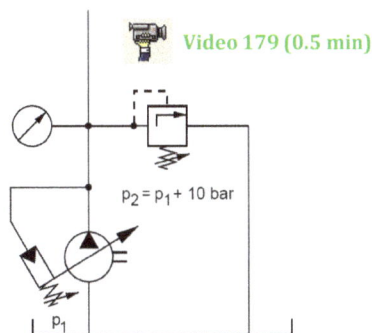

Video 179 (0.5 min)

$p_2 = p_1 + 10$ bar

p_1

Fig. 4.3- Pressure Compensated Pump with a Backup Pressure Relief Valve Overrunning Load

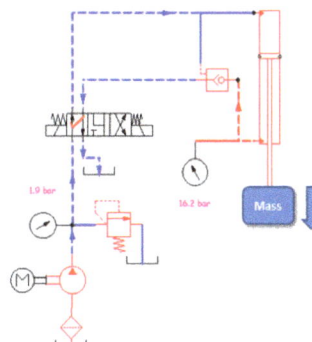

Fig. 4.4- Pilot Operated Check Valve to Control Overrunning Load

13

13

4.2.8- Excessive Pressure at the Pump Outlet

T-Pump-08-Excessive Pressure at the Pump Outlet	
Pressure relief valve is set too high.	• Reset the valve.
Is it a variable displacement pump?	• Check the setting of the pump controller.
External load is above normal?	• Check load conditions.
Pressure relief valve is at fault.	• Consult Chart: • "T-Valve-03-PCV Troubleshooting".

**Table 4.9- Troubleshooting Chart
(T-Pump-08-Excessive Pressure at the Pump Outlet)**

14

14

4.2.9- Leaking Pump

T-Pump-09-Leaking Pump	
Pump is very hot?	• Consult Charts: • "T-Unit-03-Excessively Hot Unit". • "T-System-04-Excessive System Heat".
Pump is over pressurized?	• Check working pressure versus max allowable pressure for the pump. • Consult Chart: "T-Pump-08: Excessive Pressure at the Pump Outlet".
Pump is noticeably noisy and vibrating?	• Consult Charts: • "T-Unit-02-Noisy Unit". • "T-Pump-12-Excessive Pump Noise and Vibration". • "T-System-03-Excessive System Noise and Vibration".

Table 4.10- Troubleshooting Chart (T-Pump-09-Leaking Pump)

15

15

Lose or broken pressure lines? Damaged thread of pump ports? Use of incorrect fittings?	▪ Tighten or replace the line. ▪ Repair/replace pump housing. ▪ Use standard fittings.
Case drain restricted or too small?	▪ Check case drain pressure. ▪ Size the drain line properly. ▪ Eliminate kinks and bends.
Pump housing cracked due to mechanical stress or vibration?	▪ Replace the pump. ▪ Tighten pump housing properly. ▪ Isolate pump housing from vibration.
Shaft seal leak? Contamination between shaft and seal?	▪ Consult Chart: ▪ **"T-Seals-01-Seal Troubleshooting".**
Excessive pump wear?	▪ Consult Charts: ▪ **"T-Pump-10-Excessive Pump Wear".**

Table 4.10 - Continue

16

16

4.2.10- Excessive Pump Wear or Inside Parts Broken

T-Pump-10-Excessive Pump Wear	
Fluid Aeration?	▪ Consult Chart: ▪ **"T-System-01-Fluid Aeration".**
Pump Cavitation?	▪ Consult Chart: ▪ **"T-System-02-Pump Cavitation".**
Abrasive dirt in the fluid?	▪ Drain and flush the system thoroughly. ▪ Replace filter element. ▪ Investigate sources of wear base on fluid analysis.
Fluid viscosity too low or too high?	▪ Check the working temperature versus recommended fluid viscosity. ▪ Act accordingly (adjust working temperature or change fluid).

Table 4.11- Troubleshooting Chart
(T-Pump-10-Excessive Pump Wear or Inside Parts Broken)

17

17

Higher water content in fluid?	▪ Investigate sources of water penetration to the system and then act accordingly.
Working pressure is above normal?	▪ Check working pressure versus max allowable pressure for the pump. ▪ Consult Chart: **"T-Pump-08: Excessive Pressure at the Pump Outlet"**.
Pump-coupling misalignment or drive belt is too tight?	▪ Check pump shaft alignment with the coupling. ▪ Check belt tension (if found). ▪ Check maximum radial and axial load on shaft.

Table 4.11- Continue

18

18

4.2.11- Air Leaks into Pump

Video 187 (0.5 min)

T-Pump-11-Air Leaks into Pump	
Is the reservoir fluid level too low?	▪ Follow the guidelines to make up the oil in the reservoir to the specified level.
Leaking fitting in the intake line?	▪ Tighten the leaking fitting on intake line.
Pump shaft seal worn or damaged?	▪ Replace the pump shaft seal.

Table 4.12- Troubleshooting Chart (T-Pump-11-Air Leaks into Pump)

19

19

4.2.12- Excessive Pump Noise and Vibration

T-Pump-12-Excessive Pump Noise and Vibration	
Pump Cavitation (is vacuum in the pump intake below recommendations)?	▪ Consult Chart: ▪ "T-System-02-Pump Cavitation".
Excessive pump wear or inside parts broken?	▪ Consult Chart: ▪ "T-System-10-Excessive Pump Wear".
Wrong direction of pump rotation?	▪ Check electrical wiring to electric motor.
Working pressure is too high?	▪ Check working pressure vs. maximum allowable pressure of the pump.
Pump housing bolts not tightened properly.	▪ Check and tighten according to manufacturer recommendations.
Noise damping cushions worn.	Check and replace if needed.
Coupling or other transmission elements are wrongly aligned or lose?	Check and resolve accordingly.
Is the noise due to pressure ripples from a positive displacement pump?	▪ Consider installing a small accumulator downstream the pump OR Shock suppressor.

Table 4.13- Troubleshooting Chart
(T-Pump-12-Excessive Pump Noise and Vibration)

20

20

4.3- Hydraulic Pumps Failure Analysis

4.3.1- Pump Failures due to Contamination

Source	Failure Frequency (%)
Contamination	80
Installation	12
Manufacturing Defects	6
Design	2

Table 4.14- Pump Failure Frequency due to Various Reasons

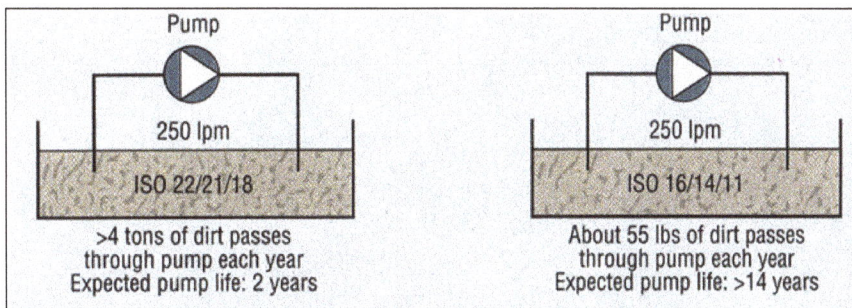

Pump

250 lpm

ISO 22/21/18

>4 tons of dirt passes through pump each year
Expected pump life: 2 years

Pump

250 lpm

ISO 16/14/11

About 55 lbs of dirt passes through pump each year
Expected pump life: >14 years

Fig. 4.5- Effect of Cleanliness Level on a Pump Lifetime
(Hydraulic & Pneumatic Magazine)

21

21

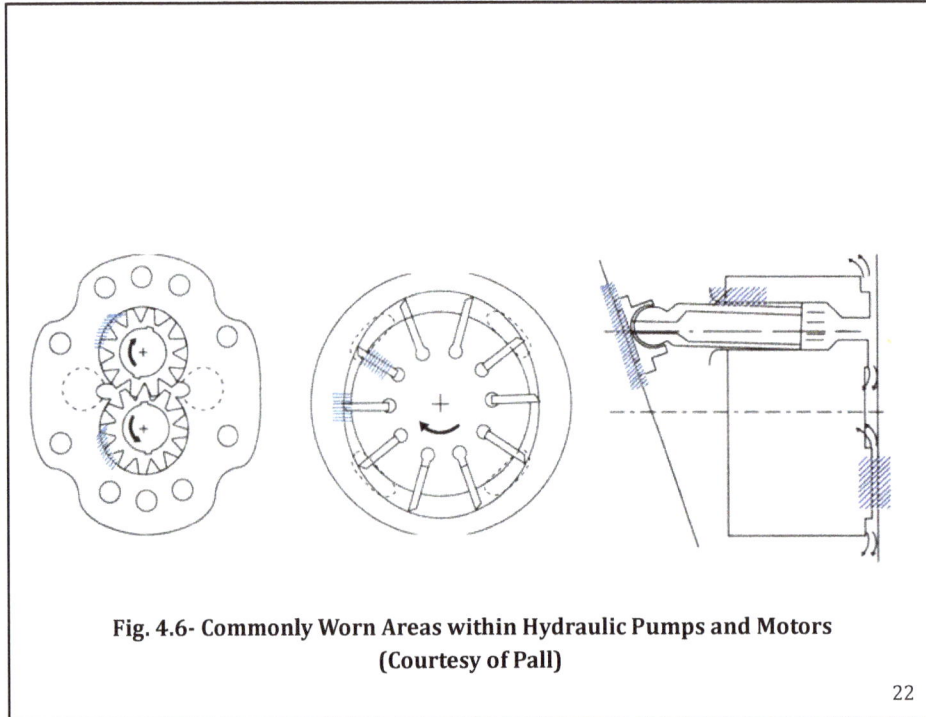

Fig. 4.6- Commonly Worn Areas within Hydraulic Pumps and Motors (Courtesy of Pall)

22

22

4.3.1.1- Failures due to Contamination in Gear Pumps

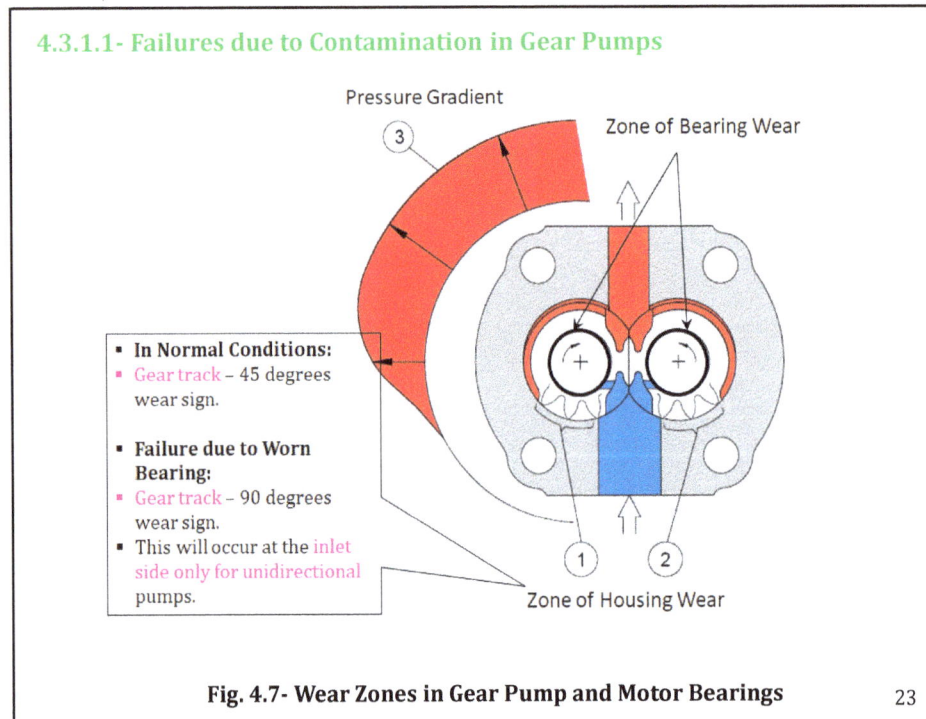

- **In Normal Conditions:**
- Gear track – 45 degrees wear sign.

- **Failure due to Worn Bearing:**
- Gear track – 90 degrees wear sign.
- This will occur at the inlet side only for unidirectional pumps.

Pressure Gradient

Zone of Bearing Wear

Zone of Housing Wear

Fig. 4.7- Wear Zones in Gear Pump and Motor Bearings

23

23

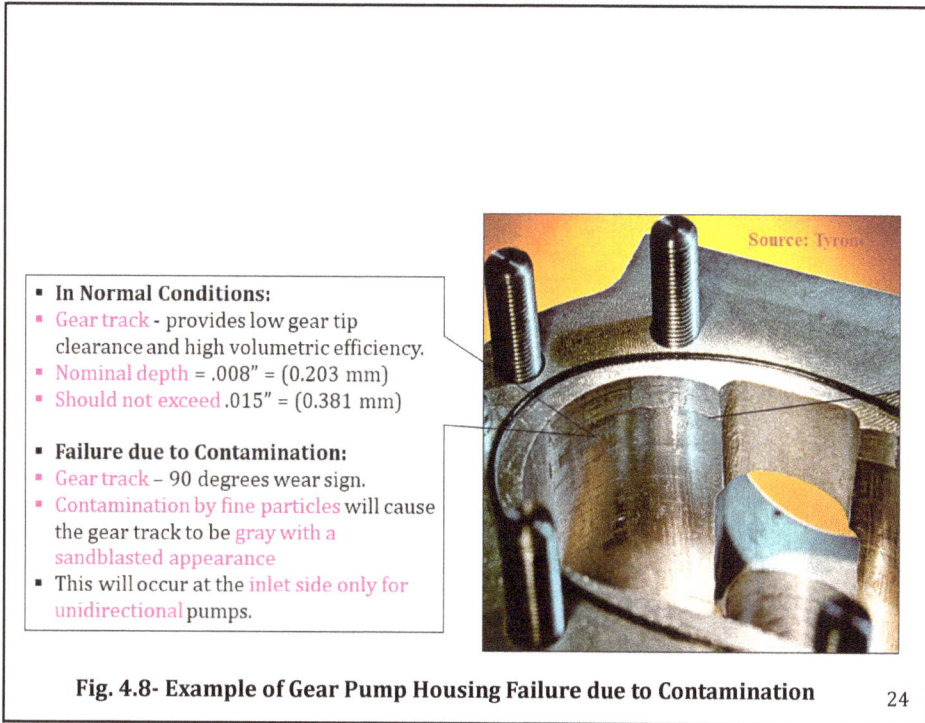

- **In Normal Conditions:**
- Gear track - provides low gear tip clearance and high volumetric efficiency.
- Nominal depth = .008" = (0.203 mm)
- Should not exceed .015" = (0.381 mm)

- **Failure due to Contamination:**
- Gear track – 90 degrees wear sign.
- Contamination by fine particles will cause the gear track to be gray with a sandblasted appearance
- This will occur at the inlet side only for unidirectional pumps.

Source: Tyrone

Fig. 4.8- Example of Gear Pump Housing Failure due to Contamination

24

24

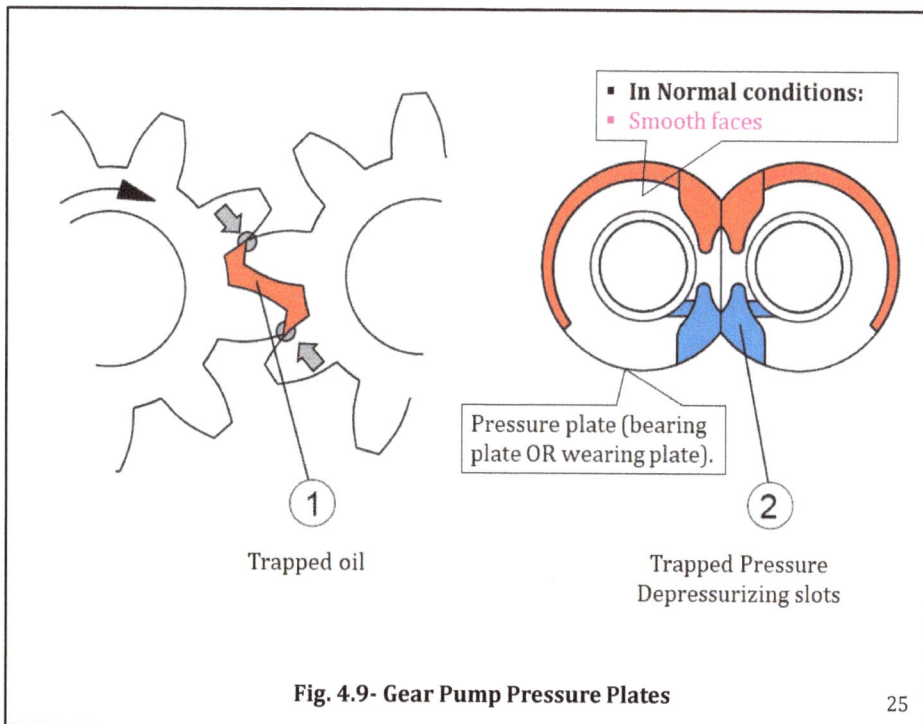

- **In Normal conditions:**
- Smooth faces

Pressure plate (bearing plate OR wearing plate).

① Trapped oil

② Trapped Pressure Depressurizing slots

Fig. 4.9- Gear Pump Pressure Plates

25

25

- **Failure due to Contamination:**
  - circular scratches caused by particles of more than 100 microns in size.

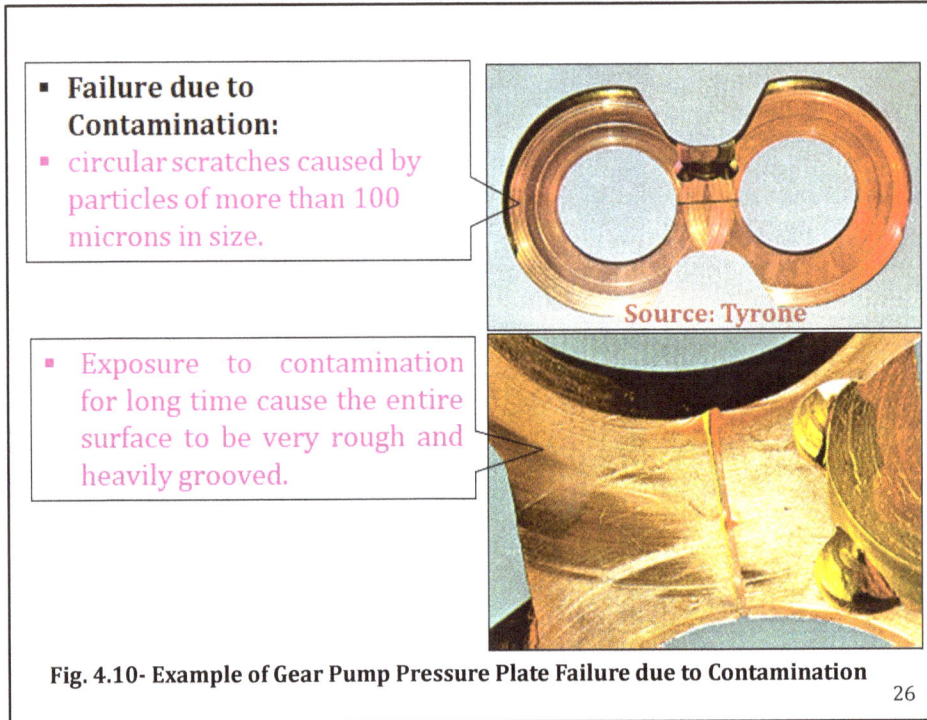

Source: Tyrone

  - Exposure to contamination for long time cause the entire surface to be very rough and heavily grooved.

Fig. 4.10- Example of Gear Pump Pressure Plate Failure due to Contamination

26

26

- **In Normal conditions:**
  - Smooth shaft.

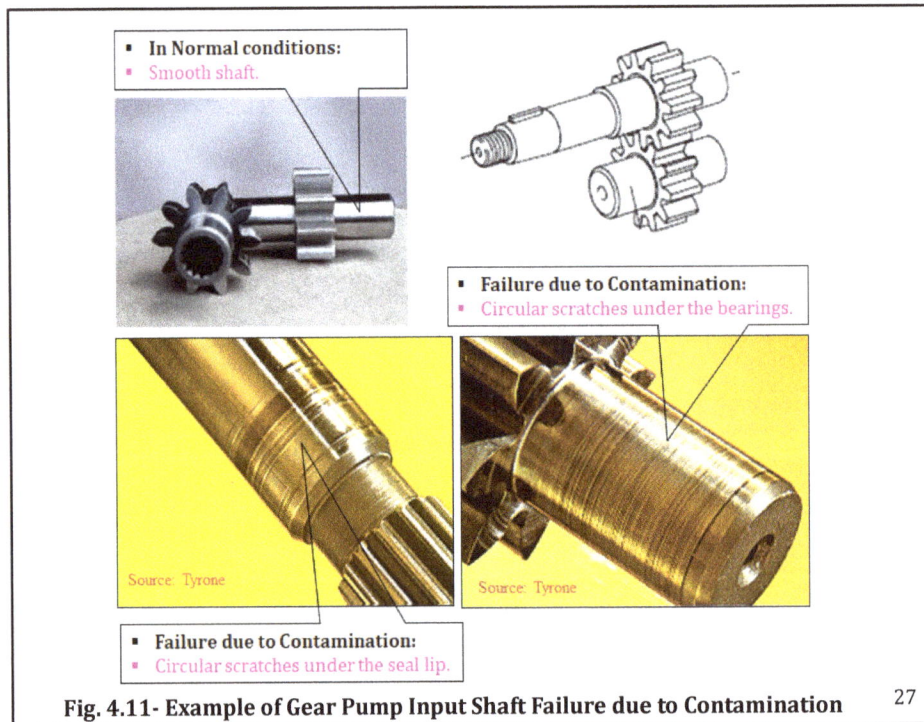

- **Failure due to Contamination:**
  - Circular scratches under the bearings.

Source: Tyrone

Source: Tyrone

- **Failure due to Contamination:**
  - Circular scratches under the seal lip.

Fig. 4.11- Example of Gear Pump Input Shaft Failure due to Contamination

27

27

1. A destroyed raceway of a ball bearing.
2. A chip embedded in a surface of an anti-friction *bearing.*
3. A destroyed roller bearing in a piston pump.

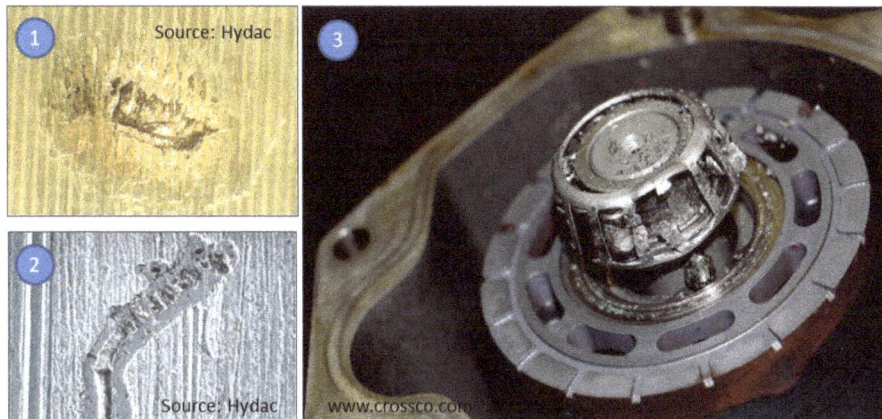

Fig. 4.12- Examples of Bearing Failures due to Particulate Contamination   28

28

4.3.1.2- Failures due to Contamination in Vane Pumps

- **This wear plate shows the damage caused by larger metal object jammed between the rotor and wear plate.**

Source: Caterpillar

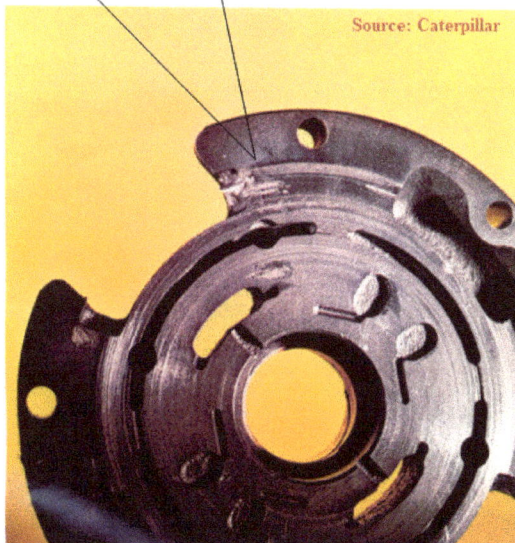

Fig. 4.13- Examples of Wear Plate Failures due Particulate Contamination   29

29

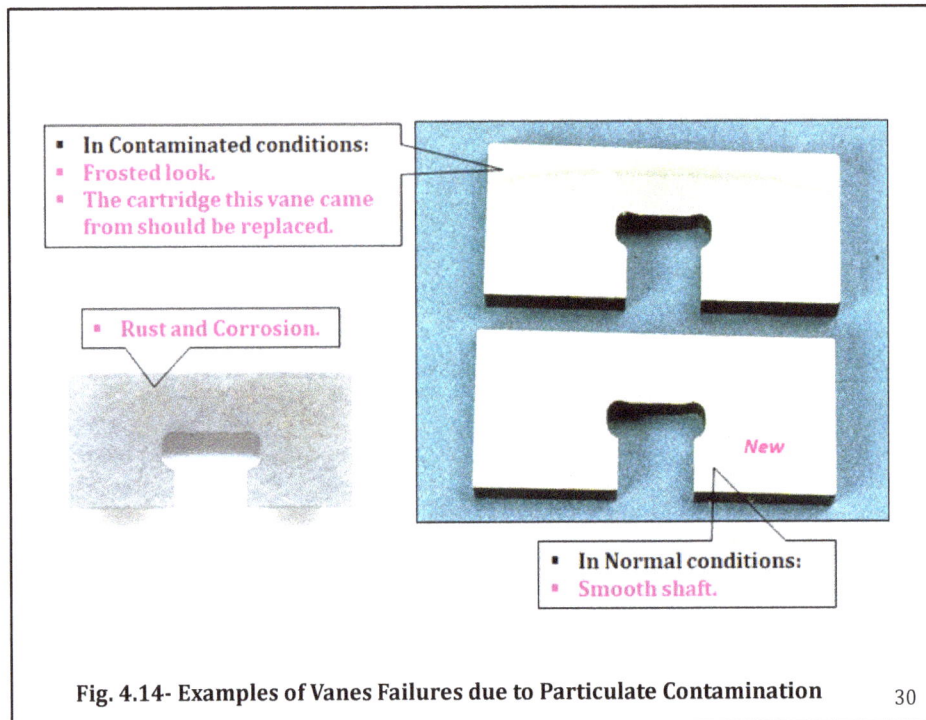

- **In Contaminated conditions:**
- Frosted look.
- The cartridge this vane came from should be replaced.

- Rust and Corrosion.

New

- **In Normal conditions:**
- Smooth shaft.

Fig. 4.14- Examples of Vanes Failures due to Particulate Contamination   30

30

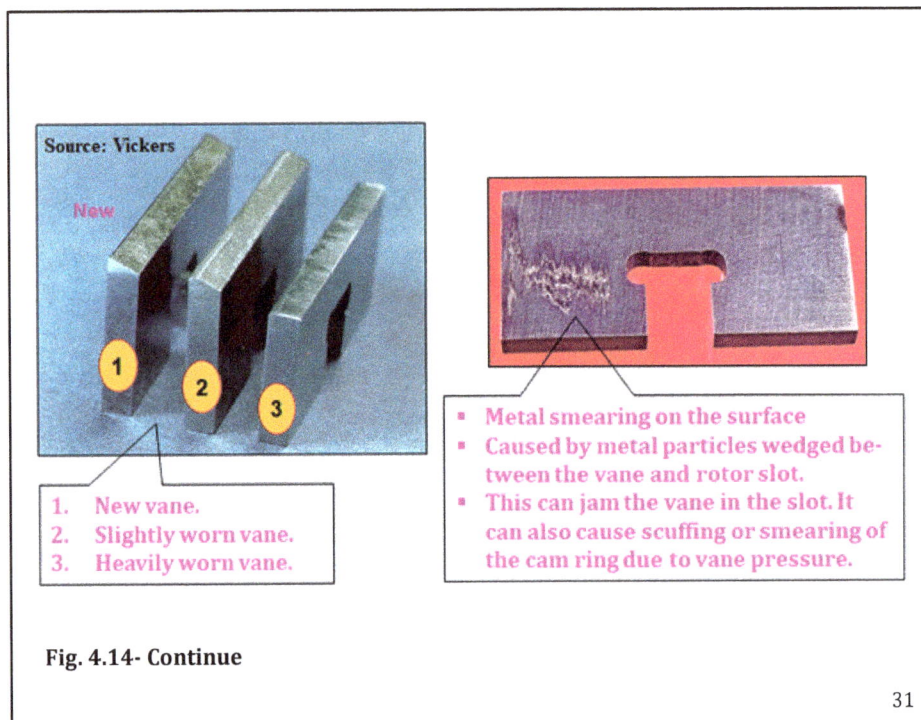

Source: Vickers

New

1   2   3

1.   New vane.
2.   Slightly worn vane.
3.   Heavily worn vane.

- Metal smearing on the surface
- Caused by metal particles wedged between the vane and rotor slot.
- This can jam the vane in the slot. It can also cause scuffing or smearing of the cam ring due to vane pressure.

Fig. 4.14- Continue

31

31

Notes:

- Worn *cam ring* in a vane pump → pump flow rate increased.
- Worn cam ring must be replaced.

Fig. 4.15- Examples of Cam Ring Failures due to Particulate Contamination 32

32

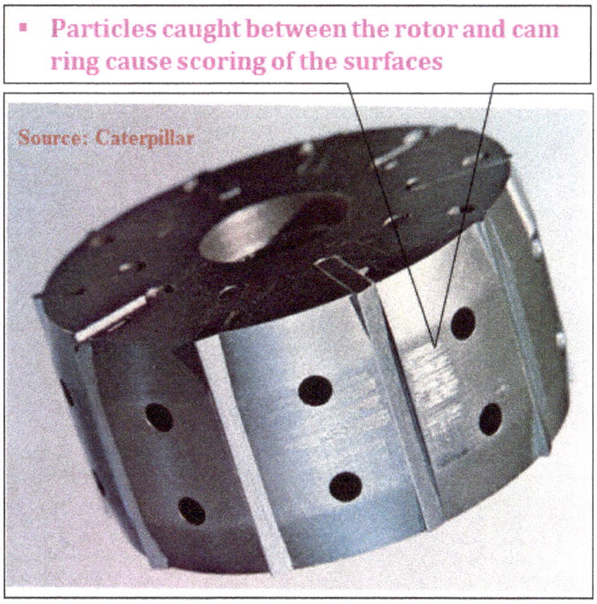

Fig. 4.16- Examples of Rotor Failures due to Particulate Contamination 33

33

4.3.1.3- Failures due to Contamination in Piston Pumps

Blocking Lubrication Passages:

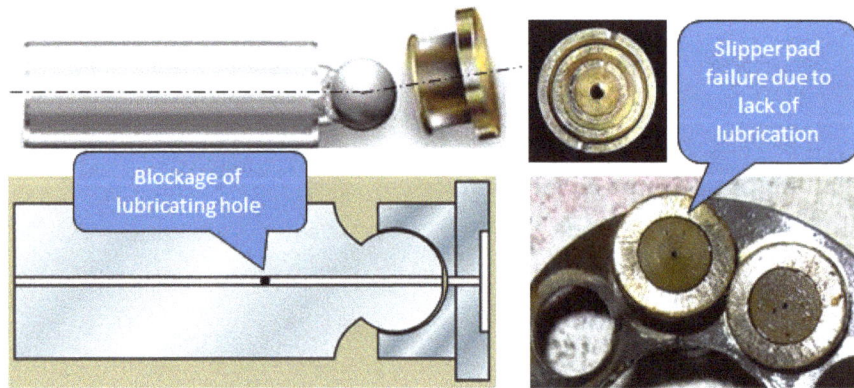

Fig. 4.17- Commonly Worn Areas within Hydraulic Pumps and Motors (Courtesy of Pall)

34

34

Cylinder Block Failure due to Contamination:
- Cylinder block top surface scoring →
- Cylinder block can be re-lapped or reground .005" to .015" (0.127 – 0.381 mm).

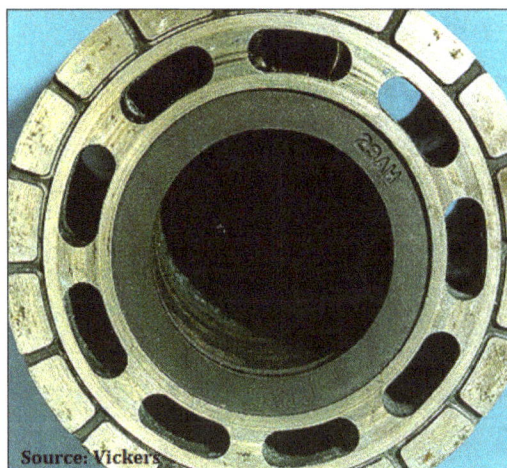

Fig. 4.18- Example of Cylinder Block Failure due to Contamination

35

35

Failure of Valve Plate due to Contamination:

- <u>Fig 4.19:</u> Valve plate surface slightly scored →
- Scored surfaces resurfaced up to .015" (0.381 mm)
- *Silencing Grooves* should not be affected.
- <u>Figure 4.20:</u> Valve plate was badly damaged → must be replaced.

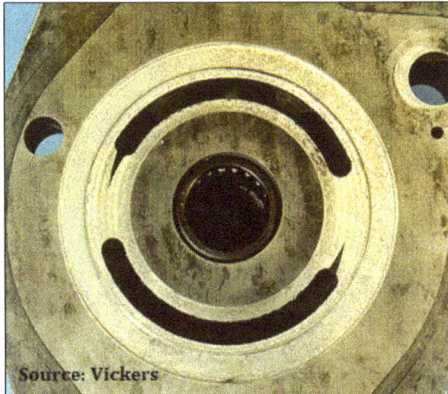

Fig. 4.19- Example of Valve Plate Failure due to Contamination

Fig. 4.20- Example of Heavily Damaged Valve Plate due to Contamination

36

36

Failure of Piston due to Contamination:

- <u>Figure 4.21:</u> damaged *piston head* → Piston head becomes loose from the shoe → old pistons should never be re-used.
- <u>Figure 4.22:</u>
  - Piston (#1) is in relatively good condition → can be reused.
  - Piston (#2) is badly damaged → must e replaced.

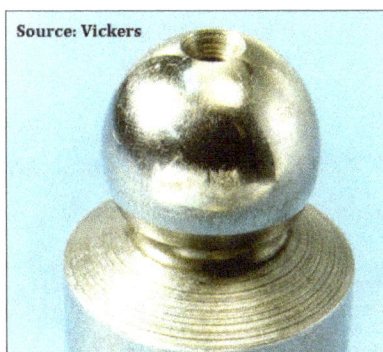

Fig. 4.21- Example of Piston Head Failure due to Contamination

Fig. 4.22- Example of Piston Body Seizure due to Contamination

37

37

Exceeding ISO 4406 Cleanliness Level:

Fig. 4.23- Examples of Piston Pumps Failure due to Particulate Contamination

38

38

4.3.2- Pump Failures due to Overheating
4.3.2.1- Failures due to Overheating in Gear Pumps

Failure of Pressure Plate due to Overheating:
- When temperature is raised above 300 ^{O}F →
- Entire pressure plate is coated with a black layer →
- Pressure plate can't be reused.

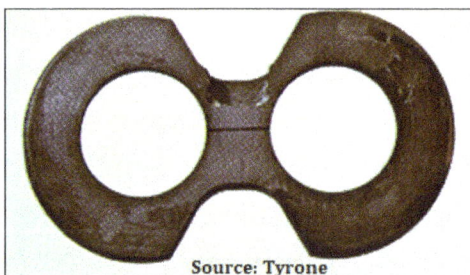

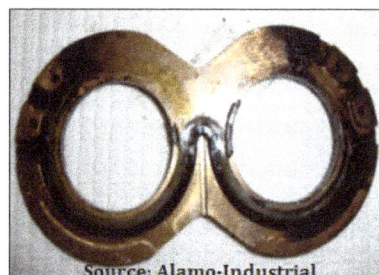

Fig. 4.24- Example of Pressure Plate Failure due to Overheating

39

39

Failure of Input Shaft due to Overheating:

- Shaft, gears, and bearing surfaces will be black all over.
- The Shaft shows some bright circles, but no grooves.
- The Gear sides near the gear face are discolored.
- Continued operation → gear face & thrust plate will start to weld together.
- → Pump seizure.
- Overheated parts shall not be used.

Video 390 (4.0 min)

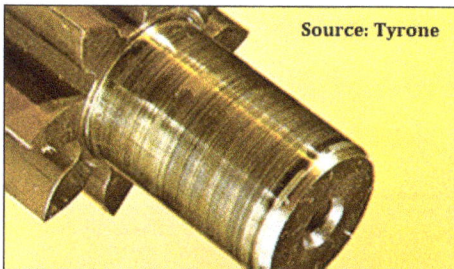

Fig. 4.25- Example of Input Shaft and Gears Failure due to Overheating

40

40

Failure of Seal Strip due to Overheating:

- Overheated *seal strip* → hardened and became brittle → Broken.

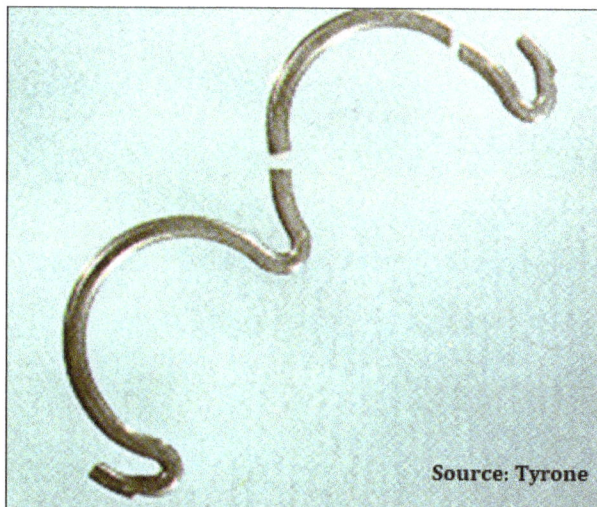

Fig. 4.26- Example of Seal Strip Failure due to Overheating

41

41

4.3.2.2- Failures due to Overheating in Vane Pumps

Failure of Cam Rings due to Overheating:

- Cam ring (#1) has mild rippling → it can be re-polished.
- Cam ring (#2) exposed to high temperature
- → vane tips literally fused into the came ring → pump seizure.
- → A ring cannot be reworked.

Fig. 4.27- Example of Cam Ring Seizure Damage due to Overheating

42

42

Failure of Cam Rings, Rotor, and Bearing due to Overheating:

- Left side:
  - Overheating → *discoloring* of a cam ring, a rotor, and a bearing.
  - Despite that no signs of cracks or grooves were shown, the entire cartridge should be replaced.
- Right side:
  - Severe overheating → lack of lubrication → dry run → seizure damage.

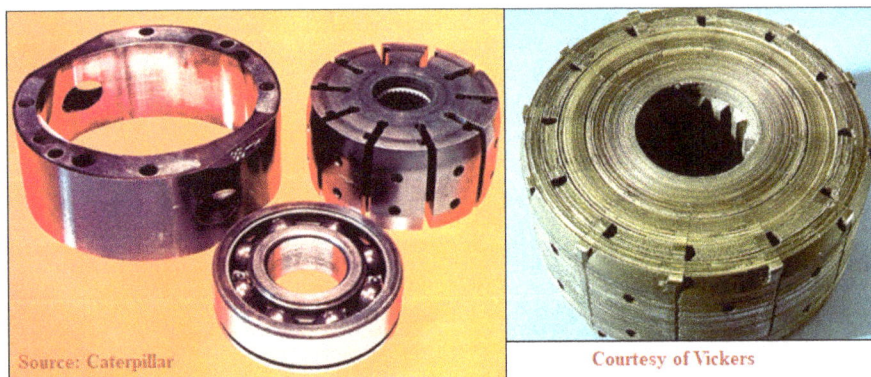

Fig. 4.28- Example of Cam Ring, Rotor, and Bearing due to Overheating

43

43

Failure of Valve Plate due to Overheating: V

- Valve plate (#1): Overheating →
- Typical amount of discoloration → no effect on the pump operation

- Valve plate (#2): Excessive overheating →
- Darkening and erosion on valve plate → such parts can't be reused.

Fig. 4.29- Example of Valve Plate Failure due to Overheating

44

44

4.3.2.3- Failures due to Overheating in Piston Pumps

Failure of Cylinder Block due to Overheating:

- A *cylinder block* is coated by black layers →
- Material properties are changed as it is heat treated →
- Cylinder block can't be reused.

Fig. 4.30- Example of Cylinder Block Failure due to Overheating

45

45

Failure of Pistons and Slipper Pads due to Overheating:

- Excessive system heat → hydraulic fluid to break down →
- Loss of lubrication → catastrophic failure to internal components.

Fig. 4.31- Example of Piston and Slipper Pad Failure due to Overheating
(Courtesy of Insane Hydraulics)

46

46

4.3.3- Pump Failures due to Cavitation
4.3.3.1- Failures due to Cavitation in Gear Pumps

Crack through bolt hole in pump body, due to severe vibration.

Sever pitting of pump housing.

Source: Caterpillar

www.advancedta.com

Fig. 4.32- Example of Gear Pump Housing Failure due to Cavitation

47

47

Failure of Gears due to Cavitation: Video 389 (2.0 min)

- Left Picture:
  o Physical wear in *gears* near the outside diameter.

- Right Picture:
  o Pump runs at high speed → cavitation → gear teeth break off → broken tooth jammed the rest of the pump → pump shaft sheared off.

Fig. 4.33- Example of Gear Failure due to Cavitation   48

48

Failure of Pump Pressure Plates due to Cavitation:

- Cavitation → collapse of bubbles → lack of oil supply →
- Discoloration & physical damage.

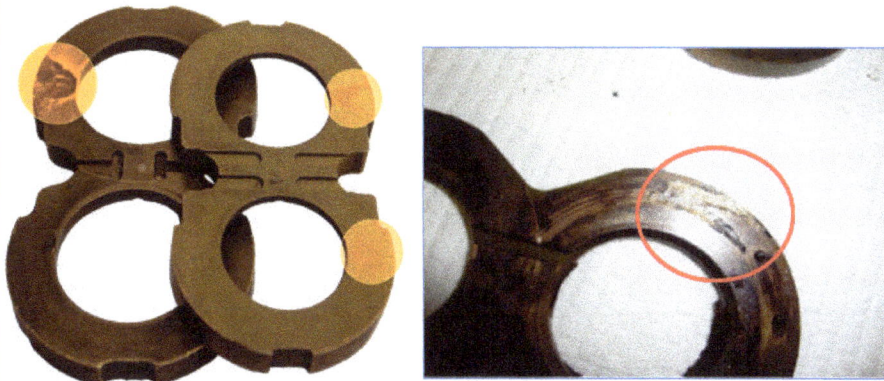

Fig. 4.34- Example of Thrust Plate Failure due to Cavitation
(www.alamo-industrial.com)   49

49

Failure of Lobe Pump due to Cavitation:

- Physical damage in housing and rotating elements.

**Fig. 4.35- Example of Failure in a Lobe Pump due to Cavitation
(www.bonvepumps.com)**

50

50

4.3.3.2- Failures due to Cavitation in Vane Pumps

Failure of Valve Plate due to Cavitation:

- Sever erosion damage is caused by collapsed air bubbles.
- This plate cannot be resurfaced.

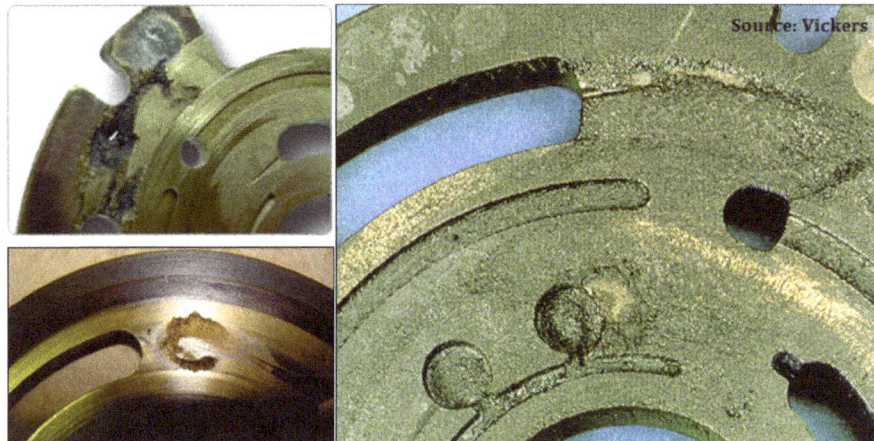

Fig. 4.36- Example of Valve Plate Failure due to Cavitation

51

51

Failure of Vanes and Cam ring due to Cavitation:

- Cam ring is severely chopped and worn.
- Vane is severely damaged.

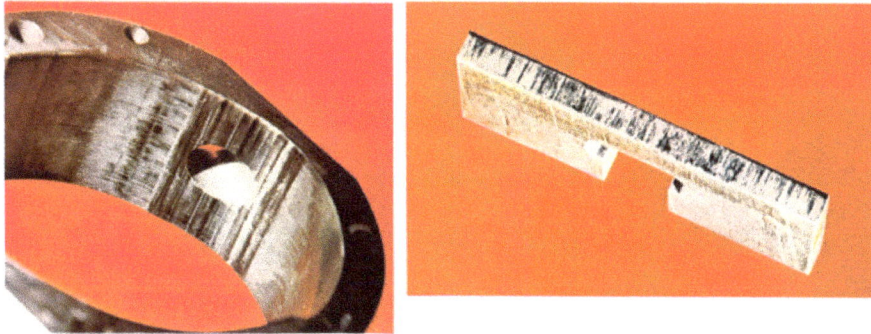

Fig. 4.37- Example of Vane and Cam Ring Failure due to Cavitation

52

52

4.3.3.3- Failures due to Cavitation in Piston Pumps

Failure of Cylinder Block due to Cavitation:

- Traces of cavitaion damages inside the *piston chambers* of a *cylinder block.*

Fig. 4.38- Example of Cavitation Damages in a Cylinder Block of a Piston Pump

53

53

Failure of Valve Plate due to Cavitation:

- Physical damage near the silencing grooves
- These *valve plates* are beyond repair.

Fig. 4.39- Example of Valve Plate Failure due to Cavitation

54

54

Slipper Pads Lift and Roll due to Cavitation:

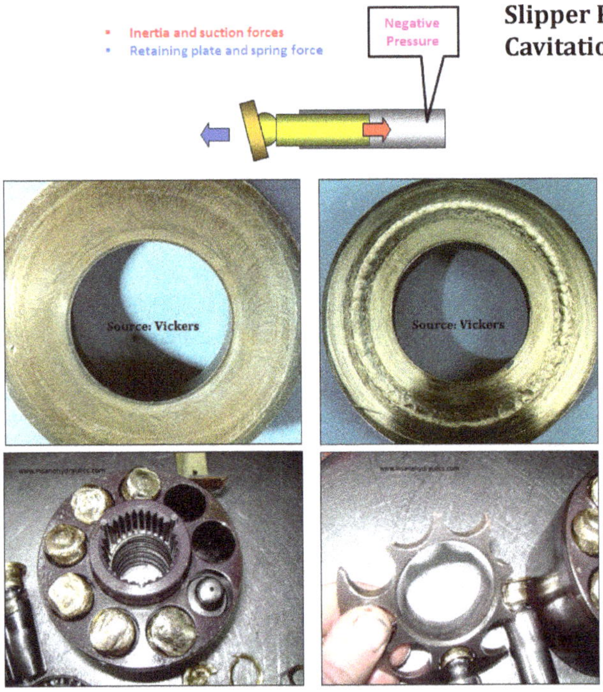

Fig. 4.40- Slipper Pads Lifting and Rolling due to Cavitation

55

55

4.3.4- Pump Failures due to Fatigue Stress

Pump Shaft Torsional Fatigue due to Pressure Spikes:

- Pressure spikes →
- Torsional fatigue →
- Shaft fracture is at random cut.

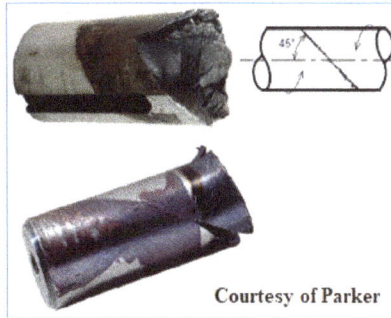

Fig. 4.41- Pump Shaft Torsional Fatigue due to Pressure Spikes

56

56

Pump Shaft Bending Fatigue due to Misalignment:

- Rotational bending fatigue →
- Shaft flex slightly with each revolution →
- Shaft is broken cleanly at a 90 deg angle to its axis of rotation.

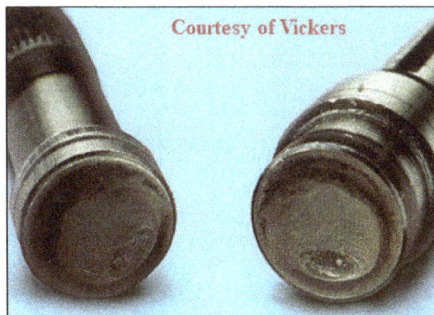

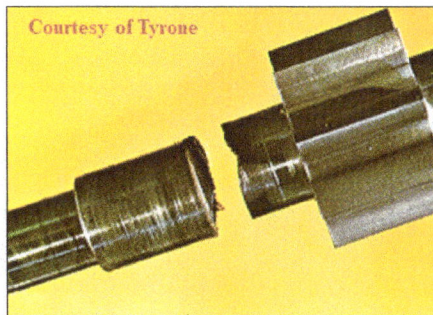

Fig. 4.42- Pump Shaft Bending Fatigue due to Misalignment

57

57

Pump Shattered Yoke in Piston Pumps:
- Variable displacement pumps →
- Pump loaded and unloaded frequently →
- Improper torqueing of bolts that secure the pins to the yoke →
- Yoke breakage.

Fig. 4.43- Yoke Fatigue due Frequent Loading

58

58

4.3.5- Pump Failures due to Overpressure

Screw Pump:
- Outlet pressure exceeded four times design value →
- Jamming of the screws → pump unexpectedly stopped →
- Shaft over-torque → shaft twist & key damage.

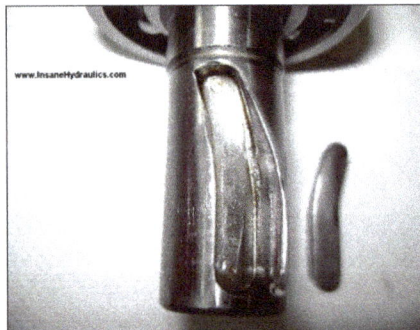

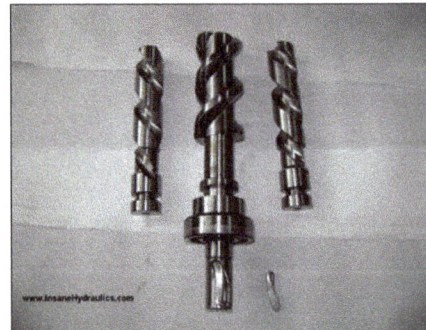

Fig. 4.44- Yoke Fatigue due Frequent Loading

59

59

Gear Pump:

- Pumps works continuously at high pressure and for extended periods →
- Stresses on the bushings that support the gear journals →
- Left: pump hosing cutout increases over the time →
- If the cutouts exceed 0.007" in depth → the pump will become inefficient
- → housing is to be replaced.
- Right: Lack of lubrication → Gears deflection → premature wear and failure.

Gear cutout in Pump Housing should not exceed .007"

Fig. 4.45- Example of Gear Pump Failures due to Overpressure
(www.alamo-industrial.com)

60

60

Vane Pump:

Left: a rotor of vane pump was ruptured.
Right: Port block was cracked due to overpressure.

Fig. 4.46- Example of Vane Pump Failures due to Overpressure
(Courtesy of Parker)

61

61

4.3.6- Pump Failures due to Insufficient Charge Pressure

- Transmission pump failure → insufficient *charge pressure* on startup →
- Cavitation on low pressure side of the hydrostatic transmission →
- Main pump failure.

Fig. 4.47- Example of Pump Failure due to Insufficient Charge Pressure on Start-Up

62

62

4.3.7- Pump Failures due to Dry Run

- Dry run → Lack of lubrication → catastrophic failure

Fig. 4.48- Example of Pump Failure due to Dry Run (Courtesy of Insane Hydraulics)

63

63

4.3.8- Pump Failures due Low Quality

- Bronze sleeve somehow detaches itself from the cylinder block.

Fig. 4.49- Example of Pump Failure due to Low Quality
(Courtesy of Insane Hydraulics)

64

64

4.3.9- Pump Failures due Oil Breakdown

- The powerpack shown in Fig. 4.50 was functioning for years.
- Hydraulic fluid didn't change for long time →
- Hydraulic fluid breakdown → oil temperature increases →
- Sludge formation → may forbidden consequences

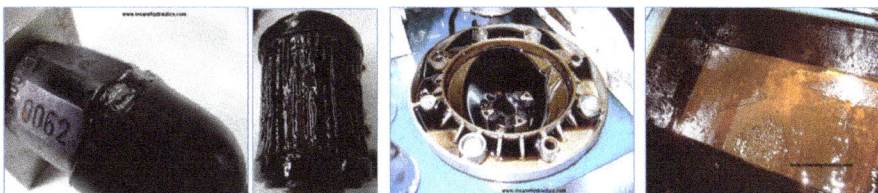

Fig. 4.50- Example of Pump Failure due to Oil Breakdown
(Courtesy of Insane Hydraulics)

65

65

4.3.10- Pump Failures due Shaft Misalignment

- Shaft misalignment → destruction of the tail and front bearing.
- **Caution:** When a machine making a little ungual noise, don't wait until the catastrophic failure point.

Video 175 (0.5 min)

Video 391 (4.0 min)

Fig. 4.51- Example of Pump Failure due to Shaft Misalignment (Courtesy of Insane Hydraulics)

66

66

4.3.11- Pump Failures due Improper Priming

- Inexperienced technical personnel involved in equipment re-commission →
- Pump isn't primed → bad start-up procedure

Fig. 4.52- Example of Pump Failure due to Improper Priming (Courtesy of Insane Hydraulics)

67

67

4.3.12- Pump Failures due Lack of Overhauling

- A pump exceeds 20,000 hours of work →
- Pump overhauling were ever performed →
- Although filters were changed at regular intervals, parts are naturally worn out.

Video 192 (2 min)

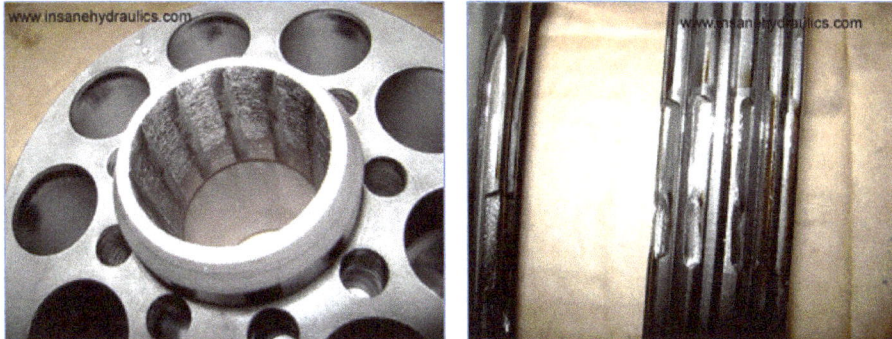

Fig. 4.53- Example of Pump Failure due to Lack of Overhauling
(Courtesy of Insane Hydraulics)

68

68

Chapter 4 Reviews

1. Which set of the following reasons leads to pump cavitation?
   A. Cold fluid, wrong fluid viscosity, fluid aeration, excessive pump wear, sticking pump parts, and low pump speed.
   B. Cold fluid, suction strainer clogged or too small, bore of suction line too small, too many bends in suction line, suction line too long, pump runs too fast.
   C. Fluid aeration, motor over-speeding and motor case drain restricted.
   D. Internal leak, air in the oil, air is not bled properly, seals sticking, pump shaft misalignment.

2. Which set of the following reasons leads to low or erratic pressure at the pump outlet?
   A. Pump runs at low speed and low pressure.
   B. Pump runs at high speed and high pressure.
   C. Pimp cavitation, fluid aeration, improper set of pump controller, worn pump, inefficient pump.
   D. Internal leak, air in the oil, air is not bled properly, seals sticking, pump shaft misalignment.

3. When no flow is showing out of the pump, this is likely because of?
   A. Pump is running too fast.
   B. Hydraulic fluid viscosity is too low.
   C. Pump-motor coupling is sheared.
   D. Pump shaft seal is leaking.

4. When no pressure is showing out of the pump, this is likely because of?
   A. Pump is running too fast.
   B. Hydraulic fluid viscosity is too low.
   C. Relief valve is set too low.
   D. Pump shaft seal is leaking.

5. When the pump experiencing excessive noise and vibration, this is likely because of?
   A. Pump is running at low speed.
   B. Hydraulic fluid viscosity is too low.
   C. Relief valve is set too low.
   D. None of the above.

6. When air is leaking into the pump, this is likely because of?
   A. Intake line isn't tightened properly.
   B. Hydraulic fluid viscosity is too low.
   C. Relief valve is set too low.
   D. None of the above.

7. In the picture shown below, a swash plate pump has the shoes of the piston rounded off. In your best guess this is because?
    A. Overpressure.
    B. Aeration of the oil.
    C. Wear due to abrasive action of contaminated oil.
    D. Lack of lubrication under one of the shoes because the lubrication hole is blocked.

8. The shown below failure in a shaft of a gear pump is likely because of?
    A. Overpressure.
    B. Pump cavitation.
    C. Hydraulic fluid particulate contamination.
    D. Overheating.

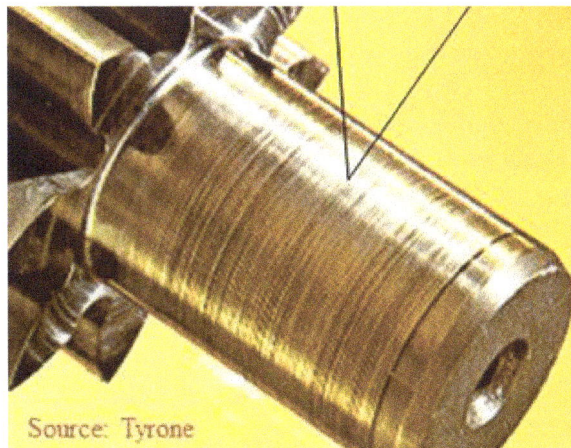

Source: Tyrone

9. The shown below failure in the cartridge of a vane pump is likely because of?
    A. Overpressure.
    B. Pump cavitation.
    C. Hydraulic fluid particulate contamination.
    D. Overheating.

Source: Caterpillar

10. The shown below failure in a valve plate of a piston pump is likely because of?
    A. Overpressure.
    B. Pump cavitation.
    C. Hydraulic fluid particulate contamination.
    D. Overheating.

Chapter 4 Assignment

Student Name: --- Student ID: ------------------

Date: -- Score: -----------------------

Question: describe the difference between pump shaft fatigue fracture due to torsional fatigue versus due to misalignment.

Chapter 5
Troubleshooting and Failure Analysis of Motors

Objectives:

This chapter discusses hydraulic *motors* inspection, troubleshooting, and failure analysis. In this chapter, a troubleshooting chart for motor faults is presented. The chapter also presents examples of defective motors due to various reasons such as contamination, clogged case drain, shaft failure, etc.

Brief Contents:

5.1- Hydraulic Motors Inspection

5.2- Hydraulic Motors Troubleshooting

5.3- Hydraulic Motors Failure Analysis

0

0

5.1- Hydraulic Motors Inspection

Hydraulic *motors* have wide varieties based on:

- **Displacement:** Fixed or Variable.
- **Rotation:** Unidirectional or Unidirectional.
- **Mechanism:** Gear, Vane, Piston, and High-Torque Low-Speed "HTLS").

Volume 1 of this series of textbooks presents good idea about the construction and operating principle of various motor mechanisms.

1

1

Hydraulic Motor Inspection Sheet	
Manufacturer	
Model #	
Serial #	
Location	
Motor Mechanism	☐ External Gear ☐ Internal Gear ☐ Gerotor ☐ Vane Motor ☐ Radial Piston [☐ Rotating Cam ☐ Rotating Cylinder Block] ☐ Bent Axis ☐ Swash Plate ☐ Other [ ]
Direction of Rotation	☐ Unidirectional ☐ Bidirectional
High Torque Low Speed Motor (HTLS)	☐ Yes ☐ NO
Motor Displacement	☐ Fixed ☐ Variable [= cc/rev]
Type of Control	☐ Pressure Compensated ☐ Displacement Controlled ☐ Constant Power (Torque) ☐ Load Sense
Drive Shaft	Type and Size:
Ports	Case Drain: ☐ Yes ☐ NO Case Drain size: Inlet Port size: Outlet Port Size:
Conditions of Seals	
Conditions of Bearings	
Conditions of Inside Parts	
Other Notes:	

Table 5.1 – Hydraulic Motors Inspection Sheet

2

2

5.2- Hydraulic Motors Troubleshooting

Many of the faults of motor mechanisms are similar to what occurs in pump mechanisms. Therefore, some of the pump troubleshooting charts will be used in motor troubleshooting. So, when any of the pump troubleshooting charts are indicated here, **it should be used as applicable for motors**.

T-Motor-01-Motor Troubleshooting	
Leaking motor?	▪ Consult Chart (as applicable for motors): ▪ **"T-Pump-09-Leaking Pump"**.
Excessive motor wear?	▪ Consult Chart (as applicable for motors): ▪ **"T-Pump-10-Excessive Pump Wear"**.
Excessive motor noise and vibration?	▪ Consult Chart (as applicable for motors): ▪ **"T-Pump-12-Excessive Pump Noise & Vibration"**.
Motor shows slow performance?	▪ Check flow received by the motor. ▪ Check controller setting for variable motors. ▪ Consult Chart: ▪ **"T-System-10-Actuator Slow Performance"**.

Table 5.2 – Hydraulic Motors Troubleshooting Chart

3

3

Motor shows fast performance?	▪ Check flow received by the motor. ▪ Check controller setting for variable motors. ▪ Consult Chart: ▪ **"T-System-11-Actuator Fast Performance"**.
Motor shows erratic performance?	▪ Check controller setting for variable motors. ▪ Check if the motor rotates below recommended minimum speed **(Note 1)**. ▪ Consult Charts: ▪ **"T-Pump-03-Erratic Flow out of the Pump"**. ▪ **"T-System-12-Actuator Erratic Performance"**.
Motor rotates in wrong direction?	▪ Consult Chart: ▪ **"T-System-13-Actuator Moves in Wrong Direction"**.
Motor fails to rotate?	▪ Consult Chart: ▪ **"T-System-14-Actuator Stops to Move"**.
Motor load drifts?	▪ Consult Chart: **"T-System-15-Actuator Load Drifts"**.
Motor leaks?	▪ Consult Chart: **"T-System-16-Actuator Leaks"**.

Table 5.2 – Continue

4

4

Note 1:

▪ If the motor is fully de-stroked→ motor speed $n_m = \infty$ (theoretically).
▪ Practically the motor stalls.

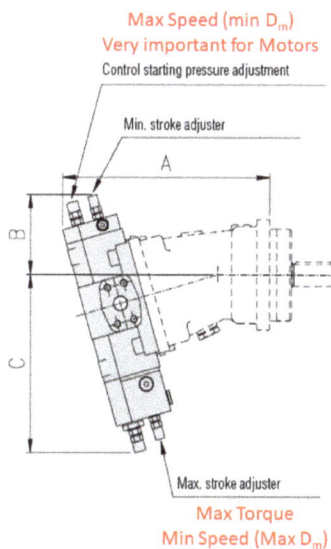

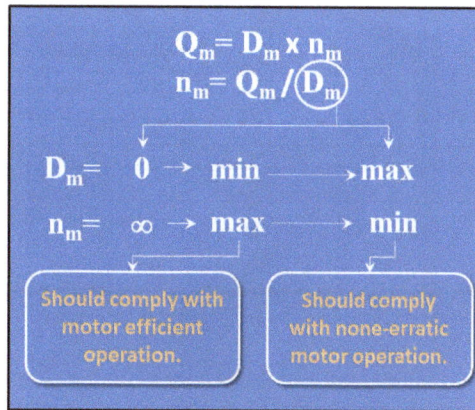

Max Speed (min D_m)
Very important for Motors
Control starting pressure adjustment
Min. stroke adjuster
A
B
C
Max. stroke adjuster
Max Torque
Min Speed (Max D_m)

$$Q_m = D_m \times n_m$$
$$n_m = Q_m / D_m$$

$$D_m = \quad 0 \rightarrow min \longrightarrow max$$
$$n_m = \quad \infty \rightarrow max \longrightarrow min$$

Should comply with motor efficient operation.

Should comply with none-erratic motor operation.

Fig. 5.1 – Recommended Speed Range of Hydraulic Motors

5

5

5.3- Hydraulic Motors Failure Analysis

- When any hydraulic motor fails → the rest of the machine stops.
- Symptoms of motors failures are similar to that in pumps.

- **Final Drive Motor Failure due to Clogged Case Drain Filter:**
- <u>Case drain filter</u> isn't checked on a regular basis →
- Case drain becomes clogged →
- Pressure will continue to build up →
- blowing seals → external leak →
- Loss of fluid →
- Severe damage can occur if the

Fig. 5.2 – Example of Hydraulic Motor Failure due to Clogged Case Drain Filter (shop.finaldriveparts.com)

6

6

Final Drive Motor Failure due to Abrasive Contaminants:
- Abrasive particles can scratch sensitive surfaces →
- Damage bearings, gear teeth, pistons → irreparable damage.

Fig. 5.3 – Example of Hydraulic Motor Failure due to Contamination (http://info.texasfinaldrive.com/)

7

7

Hydraulic Motor Shaft Failure due to Overload:

- *Shock* loads and sudden *pressure spikes* → pump and motor shaft failure.
1. Transverse torsional shear due to a high, single overload application.
2. Torsional overload resulting in transverse shear with swirl pattern.

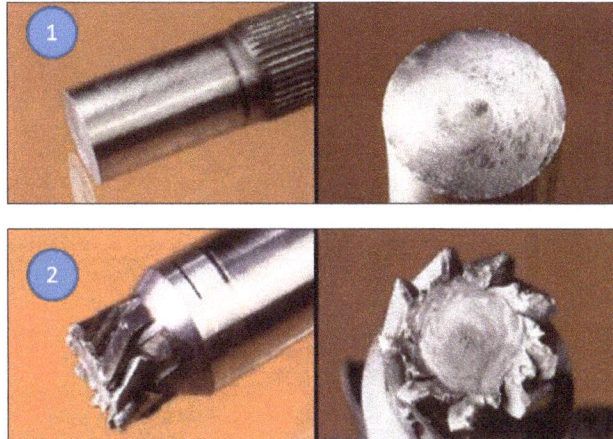

Fig. 5.4 – Example of Hydraulic Motor Shaft Failure due to Overload
(www.alamo-industrial.com)

8

8

Gear Motor Failure due to Over-torqueing:

Motor torque exceeds the rated value → one or more of the following occur

- Leak in the shaft seal area.
- front flange was "slightly worn".
- The spline coupling that nice crack in it (1),
- The shaft key was sheared (2).
- transmission was secured by the "friction weld" regime.

Fig. 5.5 – Example of Hydraulic Motor Failure due to Overload
(Courtesy of Insane Hydraulics)

9

9

Piston Motor Failure due to Contamination:

- Oil contamination → rotary group was completely destroyed.

Fig. 5.6 – Example of Hydraulic Motor Failure due to Oil Contamination (Courtesy of Insane Hydraulics)

10

10

Chapter 5 Reviews

1. If a hydraulic motor is leaking, this is likely because of?
   A. Case drain is restricted or too small.
   B. Motor works at high than normal pressure.
   C. Contamination build up between motor shaft and shaft seal.
   D. All of the above.

2. If a hydraulic motor is running slower than normal, this is likely because of?
   A. Hydraulic fluid is too hot so that viscosity is drastically decreased and internal leakage through the motor is accordingly increases.
   B. Motor works at high than normal pressure.
   C. Motor drives a load that is lighter than normal.
   D. Motor is used in a closed hydraulic circuit.

3. If a hydraulic motor stops to move, this is likely because of?
   A. Motor shaft seal is leaking.
   B. No pressure builds up at the pump outlet.
   C. Motor drives a load that is lighter than normal.
   D. Motor is used in an open hydraulic circuit.

4. The shown below failure in a motor shaft, this is likely because of?
   A. Motor overpressure/over-torqueing.
   B. Motor overspeed.
   C. Motor overheating.
   D. Hydraulic fluid particulate contamination.

5. The shown below failure in a motor shaft, this is likely because of?
   A. Motor overpressure/over-torqueing.
   B. Motor overspeed.
   C. Motor overheating.
   D. Hydraulic fluid particulate contamination.

Chapter 5 Assignment

Student Name: -- Student ID: ------------------

Date: --- Score: -----------------------

Question: Explain, why decreasing the displacement of a variable displacement motor of a special concern and requires special attention?

Chapter 6
Troubleshooting and Failure Analysis of Cylinders

Objectives:

This chapter discusses hydraulic *cylinders* inspection, troubleshooting, and failure analysis. In this chapter, a troubleshooting chart for cylinder faults is presented. The chapter also presents examples of defective cylinder due to various reasons such as contamination, improper mounting, improper load attachment, side loading, overpressure, overheating, fluid incompatibility, saltwater, external leakage, etc.

0

0

Brief Contents:

6.1- Hydraulic Cylinders Inspection

6.2- Hydraulic Cylinders Troubleshooting

6.3- Hydraulic Cylinders Failure Analysis

1

1

6.1- Hydraulic Cylinders Inspection

- Hydraulic *cylinders* could be of a tie-rod or mill type configuration.
- Volume 1 of this series of textbooks presents good idea about the construction and operating principles of hydraulic cylinders.

Tie-Rod Cylinders   **Mill-Type Cylinders**

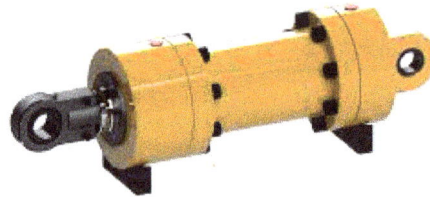

2

2

Hydraulic Cylinder Inspection Sheet	
Manufacturer	
Model #	
Serial #	
Location	
Cylinder Configuration	☐ Single Acting [☐Spring Return ☐ Load Return] ☐ Double Acting Differential [☐Single Rod ☐ Double Rods] ☐ Double Acting Synchronous ☐ Telescopic [☐ Single Acting ☐ Double Acting] Body ___ [☐ Tie Rod ☐ Mil Type] Cushions ___ [☐ Yes ☐ No]
Dimensions	☐ Bore (___) ☐ Rod (___) ☐ Stroke (___)
Ports	Port size:
Operation	☐ Push ☐ Pull ☐ both
Mounting Type	☐ Front Flange ☐ Rear Flange ☐ Pin Eye ☐ Trunnion
Position	☐ Vertical ☐ Horizontal ☐ Angle
Conditions of Rod & Load attachment	
Conditions of Rod Seal	
Conditions of Piston	
Conditions of Piston Seal	
Condition of Barrel Inside surface	
Conditions of Cyl. Cap	
Conditions of Cyl. Head	
Conditions of Ports	
Other Notes:	**Table 6.1 – Hydraulic Cylinders Inspection Sheet**

3

3

6.2- Hydraulic Cylinders Troubleshooting

T-Cylinder-01: Cylinder Troubleshooting	
External leakage from rod seals?	▪ Check cylinder rod for scoring, galling, and/or bending. ▪ Check fluid cleanliness. ▪ Consult Chart: **"T-Seal-01-Seal Troubleshooting"**.
External leakage from between the barrel and the end caps?	▪ Pressure too high. ▪ Check the tie rod torque. ▪ Check/replace static seal between barrel and end caps. ▪ Consult Chart: **"T-Seal-01-Seal Troubleshooting"**.
Cylinder internal leaking? 　Video 197 (0.5 min)	▪ Check if the cylinder is over pressurized. ▪ Check piston seal deterioration. ▪ Consult Chart: **"T-Seal-01-Seal Troubleshooting"**.
Scored cylinder rod?	▪ Replace/cover the rod **(See Note 1)**.
Bent cylinder rod? Bushing wear? Galling piton rod?	▪ Check load misalignment with the load. ▪ Check maximum allowable bending and replace the rod if needed **(See Note 2)**. ▪ Check side loads on the rod and consider using stop tubes if needed **(See Note 3)**. ▪ Check cylinder buckling due to exceeding maximum compressive load **(See Note 4)**. ▪ Check improper mounting to the machine body or load attachment.

Table 6.2 – Hydraulic Cylinders Troubleshooting Chart

4

4

Cylinder shows slow performance? 　Video 198 (0.5 min)	▪ Check flow received by the cylinder. ▪ Consult Chart: ▪ **"T-System-10-Actuator Slow Performance"**.
Cylinder shows fast performance?	▪ Check flow received by the cylinder. ▪ Consult Chart: ▪ **"T-System-11-Actuator Fast Performance"**.
Cylinder shows erratic performance? 　Video 178 (0.5 min)	▪ Consult Charts: ▪ **"T-Pump-03-Erratic Flow out of the Pump"**. ▪ **"T-System-12-Actuator Erratic Performance"**.
Cylinder moves in wrong direction?	▪ Consult Chart: ▪ **"T-System-13-Actuator Moves in Wrong Direction"**.
Cylinder stops to move?	▪ Consult Chart: **"T-System-14-Actuator Stops to Move"**.
Hydraulic cylinder has no cushioning effect/moves hard into the end position?	▪ The end position cushioning setting does not comply with the requirements

Table 6.2 – Continue

Sound 07
Hydraulic cylinder without cushioning

5

Note 1 (Covering the Rod):
- Harsh work environment (welding area and highly contaminated and dusty environment) →
- Cylinder rod must be covered

Fig. 6.1 – Protecting Cylinder Rod from Harsh Environment

Note 2 (Check for Rod Bending):
- Cylinder rod must be checked for *straightness*.
- Based on the cylinder length there is a maximum allowable bend value.

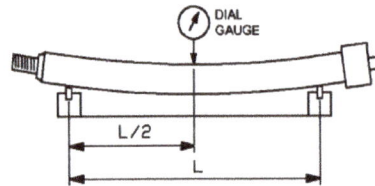

Fig. 6.2 – Checking Rod Straightness

6

6

Note 3 (Using Stop Tube):
- Piston acting as a fulcrum at one end →
- Side load rotates the rod about the piston →
- Rod bearing and rod seals failure →
- Use of *Stop Tube* reduce the reaction.

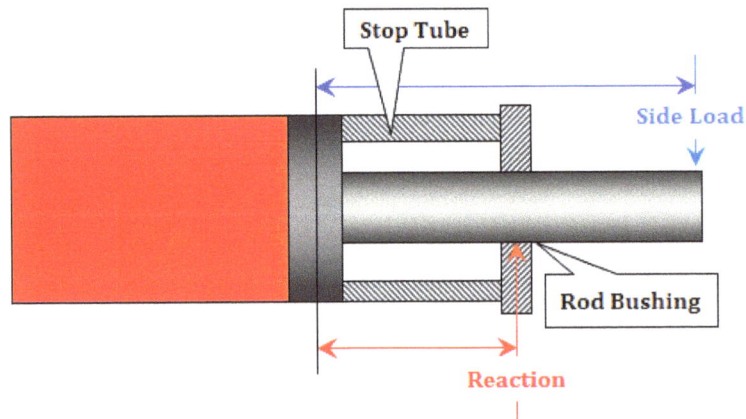

Fig. 6.3 – Use of Stop Tube to Protect Rod Bushing

7

7

Note 4 (Check for Buckling):

- Cylinder rod buckling is function of (cylinder dimensions, material, compressive load, and the way the cylinder is attached to the load).
- Buckling → destruction of rod bushing and seals → external leakage.

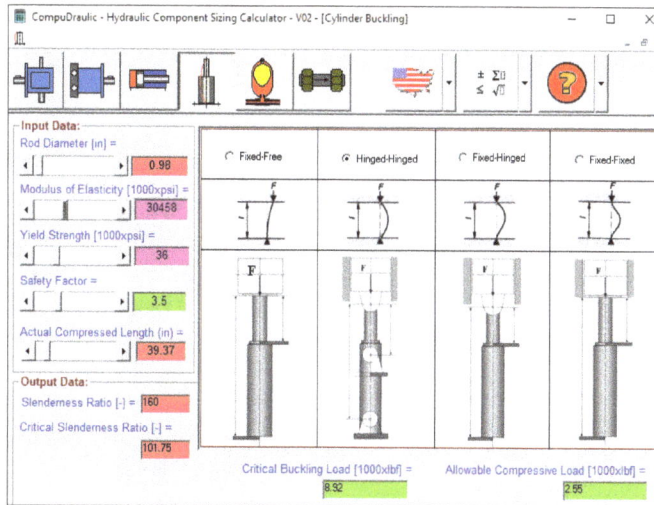

Fig. 6.4 – Cylinder Rod Buckling

8

8

6.3- Hydraulic Cylinders Failure Analysis

6.3.1- Cylinder Failure due to Particulate Contamination

Common Worn Areas due to Contamination:

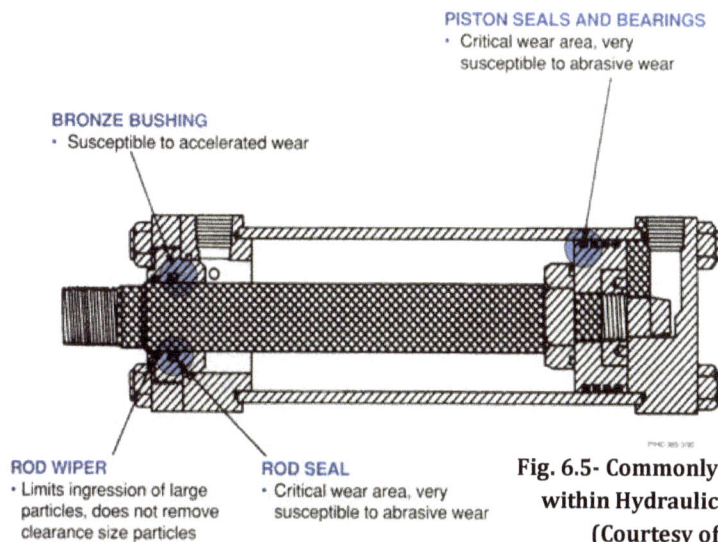

PISTON SEALS AND BEARINGS
- Critical wear area, very susceptible to abrasive wear

BRONZE BUSHING
- Susceptible to accelerated wear

ROD WIPER
- Limits ingression of large particles, does not remove clearance size particles

ROD SEAL
- Critical wear area, very susceptible to abrasive wear

Fig. 6.5- Commonly Worn Areas within Hydraulic Cylinders (Courtesy of Pall)

9

9

Cylinder Failures due to Contamination:

- Abrasive contaminants →
- seal failure →
- visible leakage.

- Abrasive contaminants →
- Piston rings were eaten away.

- Abrasive contaminants →
- scored cushion bushing →
- loss of cushioning effect.

Video 186 (0.5 min)

Fig. 6.6- Examples of Hydraulic Cylinder Failures due to Particulate Contamination

10

10

6.3.2- Cylinder Failure due to Improper Mounting

- Improper mounting → rod *bending*, and structural damage.

- For proper installation, review volume 5.

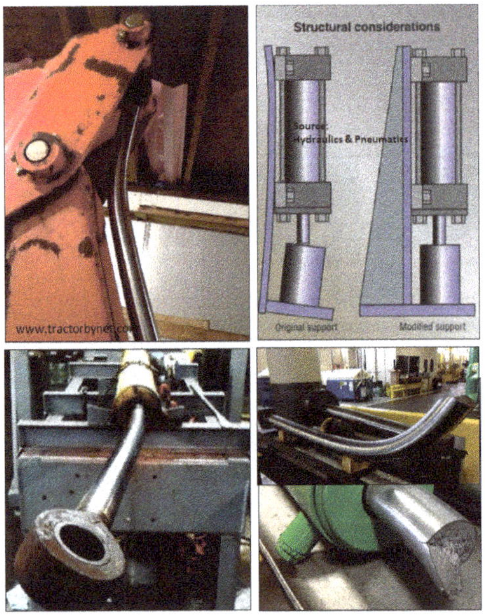

Fig. 6.7- Examples of Hydraulic Cylinder Failures due to Improper Mounting

11

11

6.3.3- Structural Failure due to Improper Load Attachment

- Upright level cylinder original assembly:
  o Rod with threaded end.
  o Hex nut to secure piston to rod.
  o Set screw to secure the hex nut to rod.

- Modification:
  o A hole was drilled through the hex nut and the rod.
  o A bolt had been inserted through the hex nut and the rod
  o The bolt was fastened with a nut.

- Result:
  o The bolt is broken into three pieces →
  o The thread of the rod assembly of the failed cylinder was ground off →
  o The rod is broken away from the cylinder barrel →
  o The aerial work platform fall in →
  o lift basket crashed to the ground (Photo courtesy of OSHA).

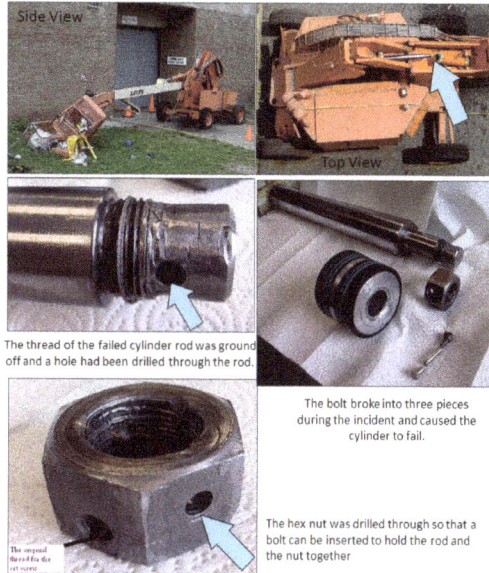

The thread of the failed cylinder rod was ground off and a hole had been drilled through the rod.

The bolt broke into three pieces during the incident and caused the cylinder to fail.

The hex nut was drilled through so that a bolt can be inserted to hold the rod and the nut together

Fig. 6.8- Example of Structural Failure due to Improper Load Attachment (metroforensics.blogspot.com) 12

12

6.3.4- Cylinder Failure due to Side Loading

Excessive Side Loading →
- Rod galling.
- Gland and seal wear.
- Brass on the seal is a mixture of gland and seal material.

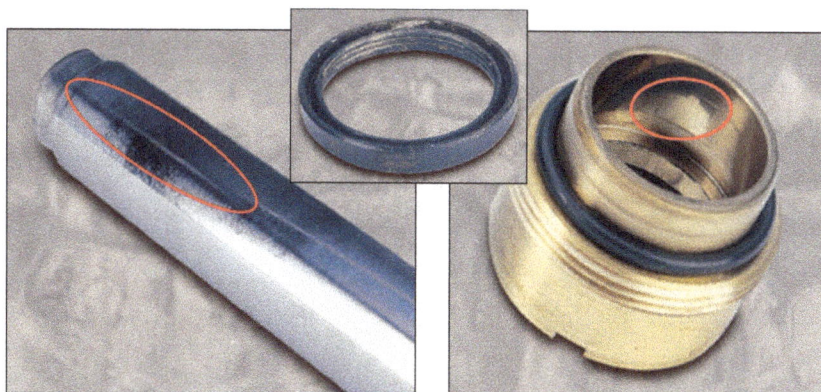

Fig. 6.9- Examples of Rod, Bushing, and Seal Failure due to Side Loads (Courtesy of Parker) 13

13

Cylinder Rod Bending due to Side Loads:

www.tractorbynet.com

Fig. 6.10- Examples of Cylinder Rod Bending due to Side Loading

14

14

6.3.5- Cylinder Failures due to Over Pressurization

Excessive working pressure →
- Cylinder burst.
- Wiper seal extrusion damage .

Video 233 (0.5 min)

Fig. 6.11- Examples of Cylinder Rod Bending due to Overpressure

15

15

6.3.6- Cylinder Seal Failures due to Over Heating

Excessive working temperature →
- Piston seal softness.
- Cracked wiper lip.

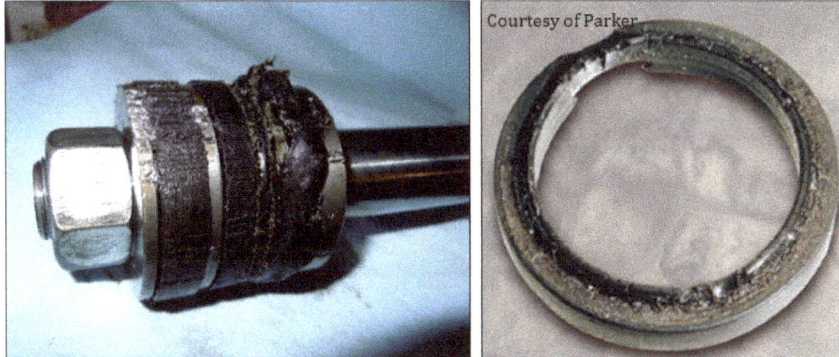

Fig. 6.12- Examples of Piston Seal and Rod Wiper Failure due to Overheating

16

16

6.3.7- Cylinder Seal Failures due to Fluid Incompatibility

New Seal   Seal damaged by water

Fig. 6.13- Examples of Piston Seal Damage due to Fluid Incompatibility

17

17

6.3.8- Cylinder Rod Corrosion due to Saltwater

- Low quality stainless-steel →
- corrosion from saltwater in marine applications →
- Pitting on cylinders → Rod seals failure →
- hydraulic fluids will leak out into the ocean.
- Such rods are unrepeatable and must be scrapped.

Fig. 6.14- Example of Cylinder Rod Corrosion due to Saltwater

18

18

6.3.9- Cylinder External Leakage

Video 603 (0.5 min)

Fig. 6.15- Examples of Cylinder External Leakage

19

19

6.3.10- Cylinder Rod Collapse due to Pressure Intensification

- Large rod diameter and meter our control →   Video 392 (5 min)
- Significant pressure intensification
- Hollow rods → cylinder rod can collapse   Video 37 (24 min)

Fig. 6.16- Examples of Cylinder Rod Collapse due to Pressure Intensification

20

20

Chapter 6 Reviews

1. The shown below seal will likely causes the cylinder to?
    A. Move faster than normal.
    B. Move slower than normal.
    C. Stop to move.
    D. Move erratically is stick-slip fashion.

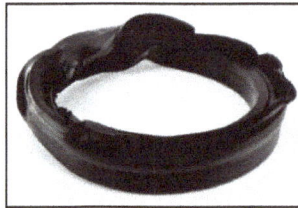

2. The shown below cylinder failure is likely because of?
    A. Overpressure.
    B. Improper cylinder mounting.
    C. Side loading.
    D. High pressure intensification at the rod side.

3. The shown below cylinder failure is likely because of?
    A. Overpressure.
    B. Improper cylinder mounting.
    C. Side loading.
    D. High pressure intensification at the rod side.

4. The shown below cylinder failure is likely because of?
    A. Overpressure.
    B. Improper cylinder mounting.
    C. Side loading.
    D. High pressure intensification at the rod side.

5. The shown below cylinder failure is likely because of?
    A. Overpressure.
    B. Improper cylinder mounting.
    C. Side loading.
    D. High pressure intensification at the rod side.

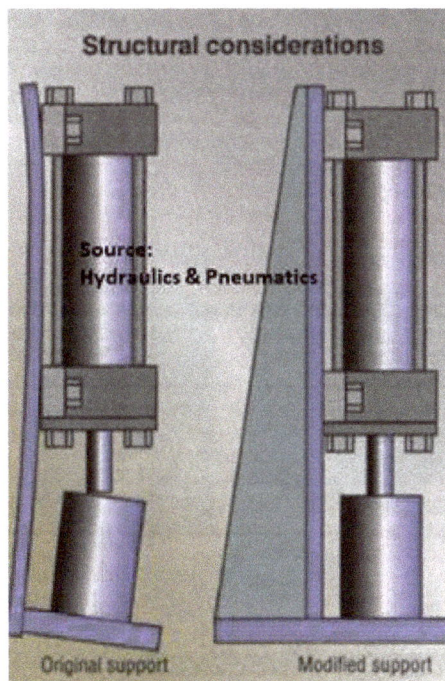

Chapter 6 Assignment

Student Name: --- Student ID: ------------------

Date: -- Score: ------------------------

Question: describe what could happed if a cylinder rod is exposed to sea saltwater.

Chapter 7
Troubleshooting and Failure Analysis of Valves

Objectives:

This chapter discusses hydraulic *valves* inspection, troubleshooting, and failure analysis. In this chapter, a troubleshooting chart for valve faults is presented. The chapter also presents examples of defective hydromechanical and electrohydraulic valves due to various reasons such as particulate and chemical contamination, solenoid burning due to inrush current, etc.

Brief Contents:

7.1- Hydraulic Valves Inspection

7.2- Hydraulic Valves Troubleshooting

7.3- Hydraulic Valves Failure Analysis

0

0

Brief Contents:

7.1- Hydraulic Valves Inspection

7.2- Hydraulic Valves Troubleshooting

7.3- Hydraulic Valves Failure Analysis

1

1

7.1- Hydraulic Valves Inspection

Hydraulic *valves* have various operating characteristics based on:

- **Valve control function:** pressure, flow, and directional.

- **Valve operation:** direct or pilot.

- **Valve control:** direct or pilot.

- **Configuration:** hydromechanical or electrohydraulic.

- **Actuation:** manual, mechanical, fluidic, and electrical.

Volume 1 of this series of textbooks presents the construction and operating principles of various valves.

2

2

Hydraulic Valve Inspection Sheet	
Manufacturer	
Model #	
Serial #	
Location	
Pressure Control Valve Type	• Direct [□Relief □Counterbalance □Sequence □Reducing] • Pilot __[□Unloading □Over-Center □ Motor Brake]
Directional Control Valve Type	• # Ports ( ) # Positions ( ) • Initial/Central Position: ( ) • Reset [□Spring □Detent] • Actuation: [□Manual □Mechanical □Pilot □Electrical] • More info ( )
Flow Control Valve Type	□Throttle □Regulator
EH Valve	Type: [□ON/OFF □Proportional □Servo] Signal: ( ) Current = Voltage = Power:
Valve Configuration	Operation: [□Direct "Single-Stage" □ Pilot "Multiple stages"] Control: _[□Direct "Internal" □ Pilot "External"] Drain: _[□ Internal □ External] Built-in Check Valve [□ Yes □ No]
Moving Element:	□ Poppet Type □ Spool Type [□Linear □Rotary]
Mounting	□ Subplate □ Line □ Manifold "Screw-In" □ Sandwich □ Other:
Ports/Flow	Port size = Rated flow Rate =
Conditions	Parts: Seals:
Other Notes:	**Table 7.1 – Hydraulic Valves Inspection Sheet**

3

7.2- Hydraulic Valves Troubleshooting

The first two actions (highlighted blue) in every valve type troubleshooting chart are common.

T-Valve-01: DCV Troubleshooting	
General Inspection	• Consult Chart: • "T-Valve-05-General Valve Troubleshooting".
Valve works electro-hydraulically?	• Consult Chart: • "T-Valve-04-EH Valve Troubleshooting".
Improper directional control operation?	• Incorrect assembly of non-identical valve spool or using the wrong spool? • Review the valve data sheet and ordering code. • Sticking valve spool
Load drifts?	• Valve null position miss-adjustment? • Follow the manufacturer's instructions about valve null adjustment.

Table 7.2– DCV Troubleshooting Chart

4

4

T-Valve-02: FCV Troubleshooting	
General Inspection	• Consult Chart: • "T-Valve-05-General Valve Troubleshooting".
Valve works electro-hydraulically?	• Consult Chart: • "T-Valve-04-EH Valve Troubleshooting".
Improper flow control Operation?	• Flow control valve may be installed backward. • Check valve size and setting. • Check pressure before and after the valve. • Plugged orifice. • Flow regulator compensator spool sticking. • Check FCV direction (Note 1).
Fluid viscosity too low or too high?	• Check the recommended range of working temperature for the valve.

Table 7.3– FCV Troubleshooting Chart

5

5

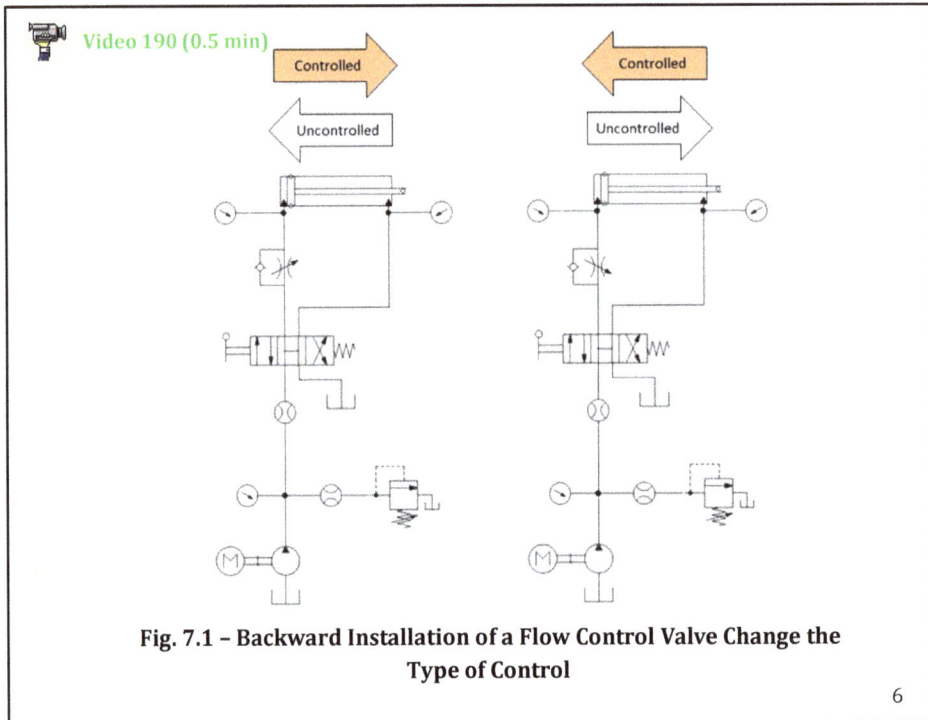

Video 190 (0.5 min)

Fig. 7.1 – Backward Installation of a Flow Control Valve Change the Type of Control

6

6

T-Valve-03: PCV Troubleshooting	
General Inspection	▪ Consult Chart: ▪ "T-Valve-05-General Valve Troubleshooting".
Valve works electro-hydraulically?	▪ Consult Chart: ▪ "T-Valve-04-EH Valve Troubleshooting".
Improper pressure control Operation?	▪ Check valve size and setting. ▪ Check for worn or broken internal parts.

Table 7.4 – PCV Troubleshooting Chart

7

7

T-Valve-04: EH Valve Troubleshooting	
Valve spool isn't moving?	▪ Use manual override **(See Notes 1)** to check whether it is a mechanical or electrical problem. ▪ Hydraulic: ▪ Valve spool is seized due to heavy *contamination*. ▪ Valve *drain* is blocked. ▪ Pilot or main spool aren't selected properly. ▪ Mechanical: ▪ Valve body is harshly distorted. ▪ Electrical: ▪ No electrical signal is received. ▪ Solenoid is burned due to low voltage, spool seizure, or high flow forces.
Valve spool response is sluggish?	▪ Hydraulic: ▪ Valve is contaminated. ▪ Valve drain is restricted. ▪ Valve body is distorted. ▪ Low pilot pressure. ▪ Electrical: ▪ Supply voltage is too low. ▪ Check setting of *Dither* signal or proportional gain. ▪ Valve is so hot so that EMF is affected **(See Notes 2)**.

Table 7.5 – EH Troubleshooting Chart

8

8

Electrohydraulic valve shuddering?	▪ Defective solenoid or *voltage* too low. ▪ Valve is undersized for the flow. ▪ Improper control settings or signal. ▪ Valve cables unshielded and affected by noise. ▪ Valve is cycled too fast **(See Notes 3)**.
External or internal electrical short?	▪ Check wiring conditions and make sure to remove reasons for possible short circuits.
Valve isn't receiving electrical control signal?	▪ Check for broken control signal wire. ▪ Check the signal generator or the controller. ▪ Check fuses or control circuit problem.
ECU of the valve isn't receiving feedback signal?	▪ Check for broken feedback signal wire. ▪ Check defective transducer, limit switch, etc.
Voltage too high or too low?	▪ Check the power supply condition vs. the nominal ratings **(See Notes 4)**.
Solenoid or torque motor of an EH valve is burned?	▪ Overlap energizing two solenoids **(See Notes 5)**. ▪ Contaminated valve cause valve seizure and burning coils due to inrush current.

Table 7.5 – Continue

9

9

Note 1 (Manual Override):
- During emergency situations, e.g. power supply failure).
- During system troubleshooting.

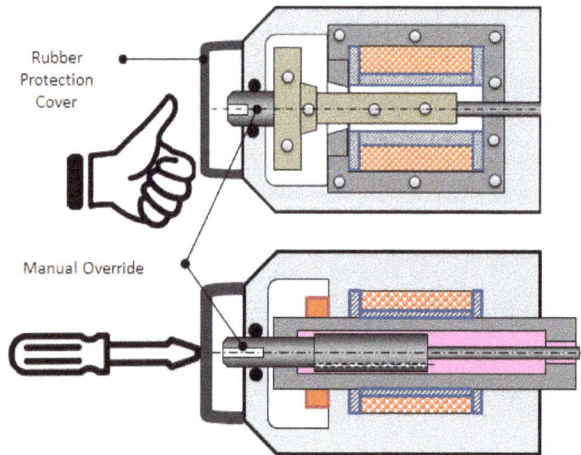

Fig. 7.2 - Manual Override

10

10

Note 2 (Solenoid Force vs. Working Temperature):
- Solenoid force is reduced due to high temperature.

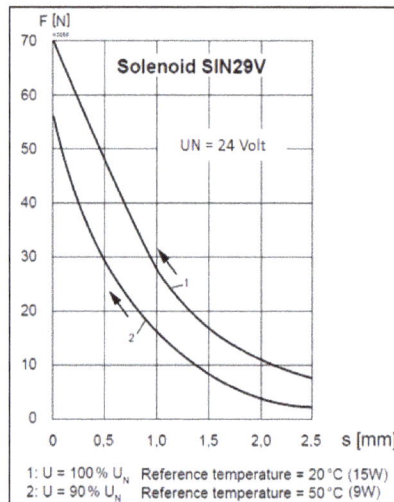

**Fig. 7.3 - Typical Force-Stroke Characteristics of a DC Switching Solenoid
(Courtesy of Wandfluh)**

11

11

Note 3 (Cycling rate of Solenoid-Operated Valves):

DC Solenoid :
- Longer switching time.
- High cycling frequency.

AC Solenoid:
- Smaller switching time
- Inrush current $\rightarrow$ longer time to cool down $\rightarrow$ lower cycling frequency.

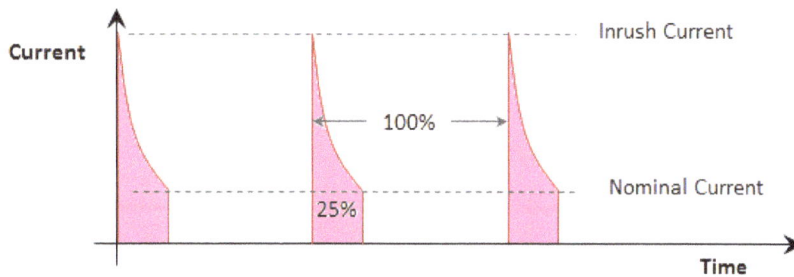

Fig. 7.4 - Switching Rate of an AC Switching Solenoid

12

12

Note 4 (EH Valve Ratings):

Nominal Voltages:
- DC (l2V, 24V, and 48V) and AC (1l0V and 220V).
- Voltage must be at least 90% of the rated current.
- High voltage heats up the coil.
- Low voltage can't seat the spool completely $\rightarrow$ high inrush current.

Nominal Currents:
- The current absorbed (1-3 Amps).

Nominal Frequencies for AC Switching solenoid:
- 60 Hz., 50 Hz, or dual frequency.

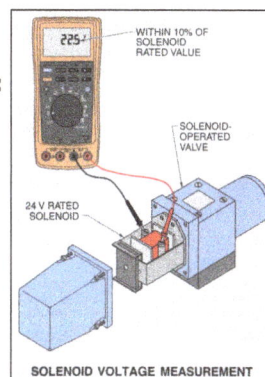

Fig. 7.5 – Measuring Voltage on Solenoid-Actuated Valve
(Courtesy of American Technical Publishers)

13

13

Note 5 (Solenoid Protection):

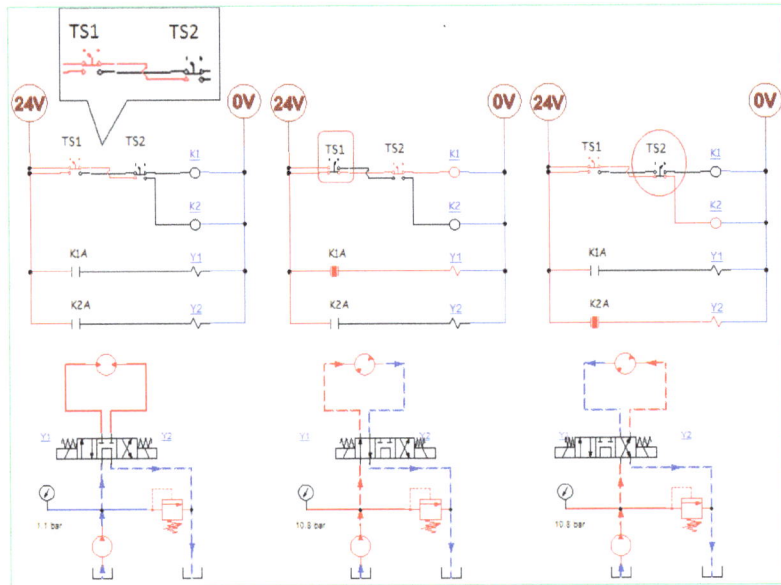

Fig. 7.6 - Mechanical Interlocking for Solenoids Protection

14

14

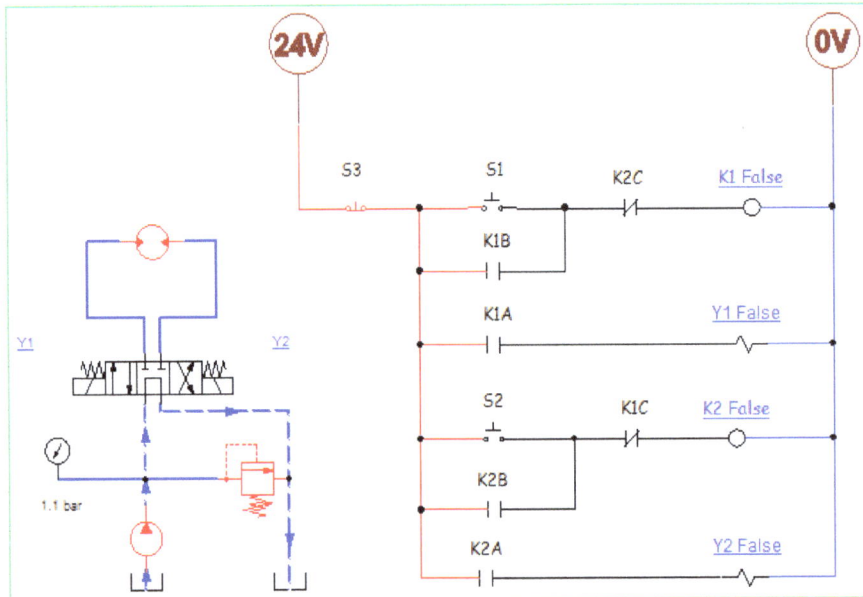

Fig. 7.7 - Electrical Interlocking for Solenoids Protection

15

15

T-Valve-05: General Valve Troubleshooting	
Valve spool isn't moving.	▪ Hydraulic: ▪ Valve spool is seized due to heavy *contamination*. ▪ Valve *drain* is blocked. ▪ Pilot operated stage isn't working properly. ▪ Mechanical: ▪ Valve body is harshly distorted.
Pilot operated stage doesn't work properly. (See Note 1).	▪ Check pilot pressure supply and value. ▪ Check the operation of the pilot valve. ▪ For internal supply of pilot pressure, make sure the central position isn't tandem or open center. Otherwise, add spring loaded check valve. ▪ Make sure pilot stage has floating center. ▪ If main spool has open or tandem center, make sure to use built in check valve. ▪ Pressure fluctuation in tank line or plugged drain line.
Valve leaks externally?	▪ Valve body distorted or cracked. ▪ Static seals failure.

Table 7.6 – General Valve Troubleshooting Chart

16

16

Valve leaks internally?	▪ Valve body distorted. ▪ Worn or broken parts. ▪ Contamination holding valve partially open. ▪ Dynamic seal failures. ▪ Valve is too hot. ▪ Valve spool scored. ▪ Leak at valve seat.
Contaminated Valve?	▪ Disassemble and clean the valve. ▪ If varnish found, investigate sources of varnish formation and act accordingly. ▪ Check the recommended cleanliness level of the valve versus the current cleanliness of the fluid.
Valve body distorted or cracked?	▪ Check if the valve is overtightened to the subplate. ▪ Realign pips connected to the valve to remove mechanical stress.
Weak or broken parts?	▪ Valve subjects to over-pressure or pressure shocks.
Valve seals failure?	▪ Consult Chart: ▪ "T-Seals-01-Seals Troubleshooting".
Valve is so hot?	▪ Consult Chart: ▪ "T-Unit-03-Excessively Hot Unit".

Table 7.6 – General Valve Troubleshooting Chart

17

17

Note 1: Basic Operating of Pilot Operated DCV:

- Internal vs External (control pressure supply and drain)

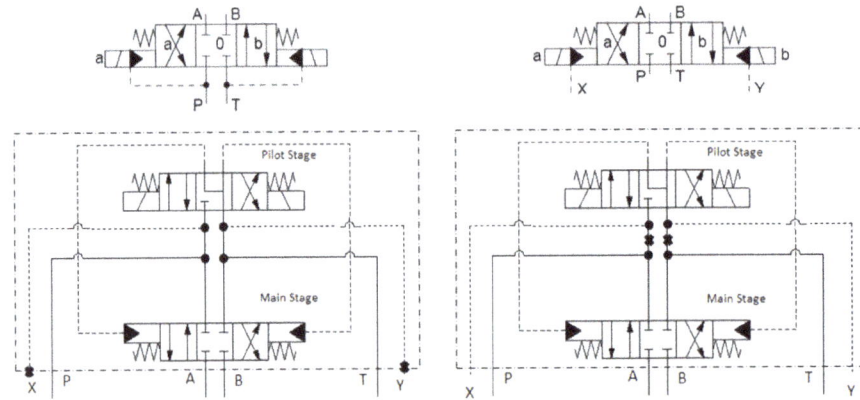

Fig. 7.8 – Pilot-Operated DCV (Courtesy of Bosch Rexroth)

18

18

Video 194 (0.5 min)

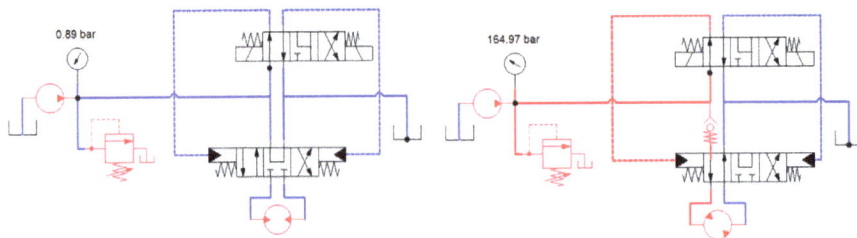

**Fig. 7.9 – Use of Spring-Loaded Check Valve
with Tandem or Open Center Main Spool**

19

19

7.3- Hydraulic Valves Failure Analysis

7.3.1- Hydraulic Valves Failure due to Particulate Contamination

Silt Lock:

- Accumulation of silt → seizure or jamming of components →
- sudden and unpredictable component failure (no wear involved) →
- Electrohydraulic valves dynamic response (stick-slip) movement.

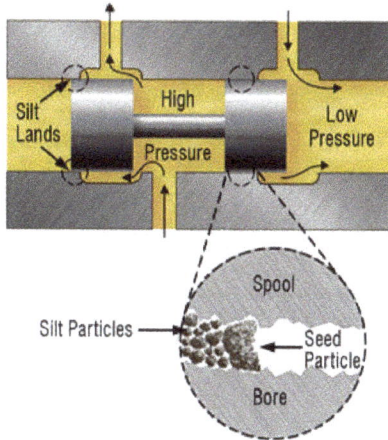

**Fig. 7.10- Silt Lock in Spool Valves
(Courtesy of Noria Corporation)**

20

20

Worn Spool Valves due to Abrasive Contaminants:

- Spools in such conditions are not reusable.

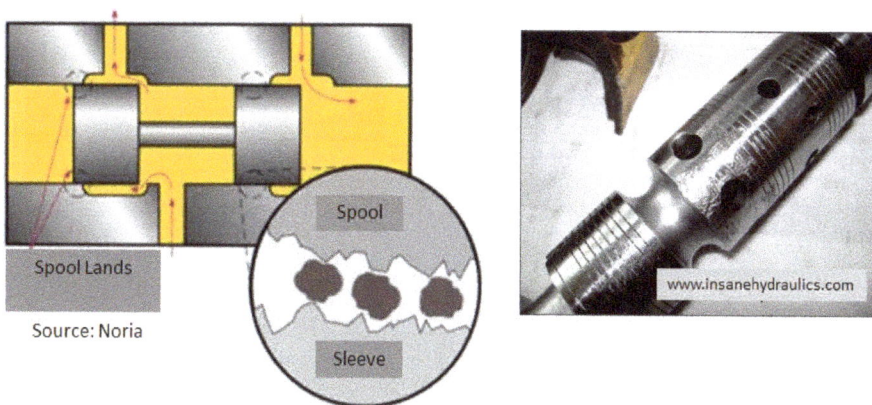

Fig. 7.11- Commonly Worn Surfaces in Spool Valves

21

21

Worn Poppet Valves due to Abrasive Contaminants:

- Valve poppet is affected by abrasive wear →
- Valve seat improperly resulting in leakage.

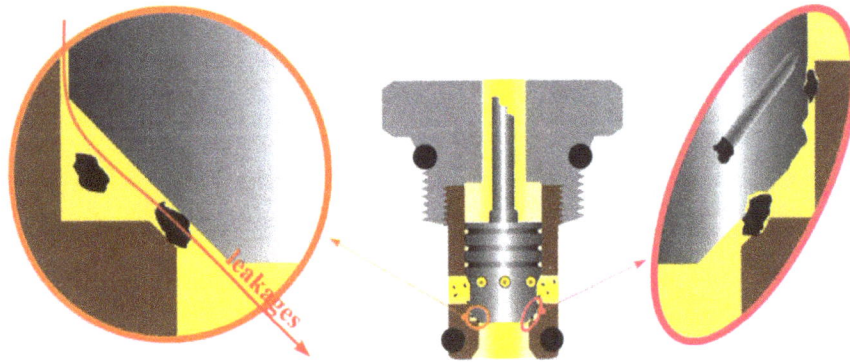

**Fig. 7.12- Commonly Worn Surfaces in Poppet Valves
(Courtesy of ASSOFLUID)**

22

22

Servo Valve Nozzle Blockage and Wear:

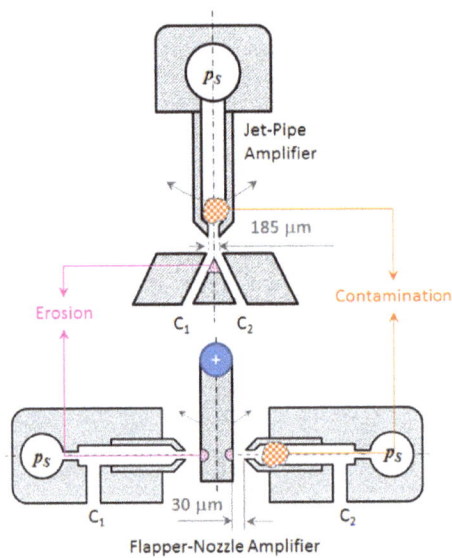

Fig. 7.13- Jet-Pipe versus Flapper-Nozzle Pilot Stage

23

23

7.3.2- Hydraulic Valves Failure due to Chemical Contamination

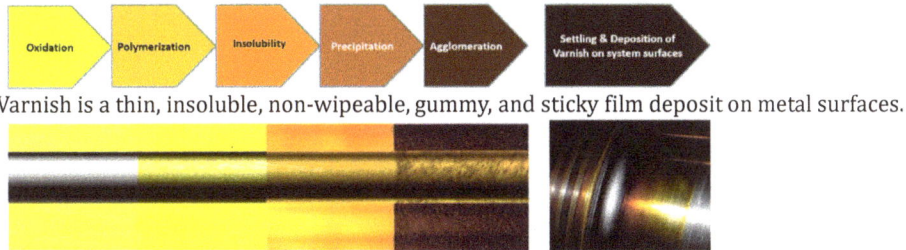

Oxidation → Polymerization → Insolubility → Precipitation → Agglomeration → Settling & Deposition of Varnish on system surfaces

Varnish is a thin, insoluble, non-wipeable, gummy, and sticky film deposit on metal surfaces.

Fig. 7.14- Varnish Formation within Hydraulic Systems (Courtesy of C.C. Jensen Inc.)

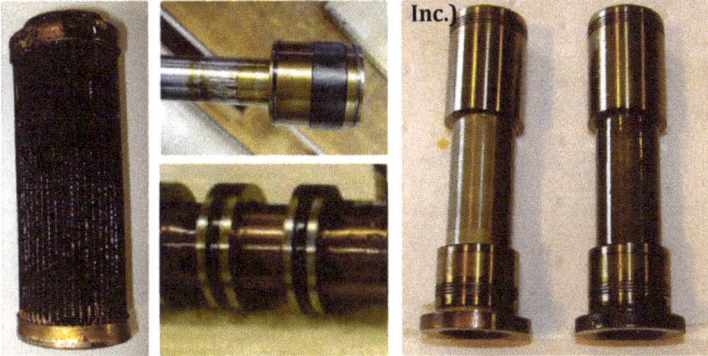

24

- Varnish layer → attracts the abrasive particles → grinding surface
- Varnish layer → fine tolerances blockage, spool seizure.

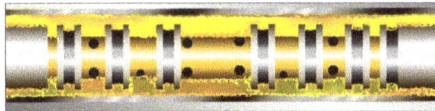

Fig. 7.15- Varnish Sticky Layer Attracts Abrasive Particles (Courtesy of C.C. Jensen Inc.)

Fig. 7.16- Varnish Sticky Layer Seizes Valve Spools (Courtesy of C.C. Jensen Inc.)

25

7.3.3- Solenoid Burn Out

- EH Valve Seizure → Inrush Current → Coil burning.

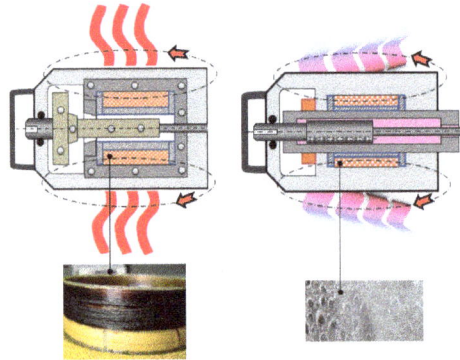

7.3.4- Spool Failure due to Dry Operation

7.3.4- Spool Failure due to Dry Operation

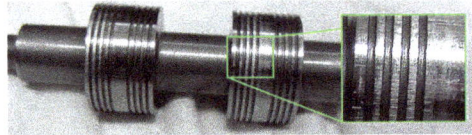

Fig. 7.18- Failed Spool due to Dry Operation (Courtesy of H & P Magazine.)

26

26

Chapter 7 Reviews

1. Which of the following set of reasons lead to a pilot operated directional valve valve spool that does not shift?
   A. Incorrect adjustment of a pressure relief valve (PRV), PRV stuck open because of contamination, worn or damaged seat of a PRV and broken spring of a PRV.
   B. Voltage is too high or too low, high ambient temperature, valve spool is blocked, two opposite solenoids energized at the same time and external or internal electrical short.
   C. Low or no pilot pressure, pilot or main valve spool sticking, valve body distortion, and solenoid of the pilot stage failed.
   D. Pre-charge is lost, pre-charge is too high, diaphragm or bladder ruptured, and pistons are seized.

2. Which of the following set of reasons lead burning out the electrical solenoids?
   A. Incorrect adjustment of a pressure relief valve (PRV), PRV stuck open because of contamination, worn or damaged seat of a PRV and broken spring of a PRV.
   B. Voltage too high or too low, high ambient temperature, valve spool is blocked, two opposite solenoids energized at the same time and external or internal electrical short.
   C. Low or no pilot pressure, pilot or main valve spool sticking, valve body distortion, and solenoid of the pilot stage failed.
   D. Pre-charge is lost, pre-charge is too high, diaphragm or bladder ruptured, and pistons are seized.

3. A servo valve is used to drive a hydraulic motor. When the servo valve receives zero input signal, the motor rotates slowly. What kind of adjustment should be made to resolve the issue?
   A. Supply of pilot pressure.
   B. Perfect spool null positioning.
   C. Viscosity of the oil used.
   D. None of the above.

4. In a closed loop electro-hydraulic system made to control the position of a hydraulic cylinder, it has been observed that the cylinder is oscillating a bit around the desired destination before it settles down. In order to resolve this issue, which of the following adjustments should do make to resolve this issue?
   A. Ramp generator setting.
   B. Decrease the proportional gain.
   C. Increase the proportional gain.
   D. Dead band eliminator adjustments.

5. In a closed loop electro-hydraulic system made to control the position of a hydraulic cylinder, it has been observed that the cylinder accelerated harshly at the beginning of both the extension and retraction stroke. In order to resolve this issue, which of the following adjustments should do make to resolve this issue?
   A. Ramp generator.
   B. Dither signal.
   C. Proportional gain.
   D. Dead band eliminator.

6. In a closed loop electro-hydraulic system made to control the position of a hydraulic cylinder, an overlapped proportional valve has been used. It has been observed that the cylinder experiencing steady state error. In order to resolve this issue, which of the following functionality you think you should enable?
   A. Ramp generator.
   B. Dither signal.
   C. Proportional gain.
   D. Dead band eliminator.

7. Steady state error can be minimized by?
   A. Increase proportional gain.
   B. Use zero-lapped spool.
   C. Enable dead-band eliminator function if an overlapped-spool is used.
   D. All the above.

8. The circuit shown below is made to perform AND function in order for a cylinder to retract automatically when both the pressure switch and the proxy switch are activated. It has been observed that, by pressing and releasing push-button S1, the cylinder is extended but never retracted. In your best understanding what would be the main reason?
   A. S2 push-button may stuck opened.
   B. 24 volt power supply is turned off.
   C. Cable break of either the pressure switch or the proxy switch.
   D. Solenoid Y1 burned out.

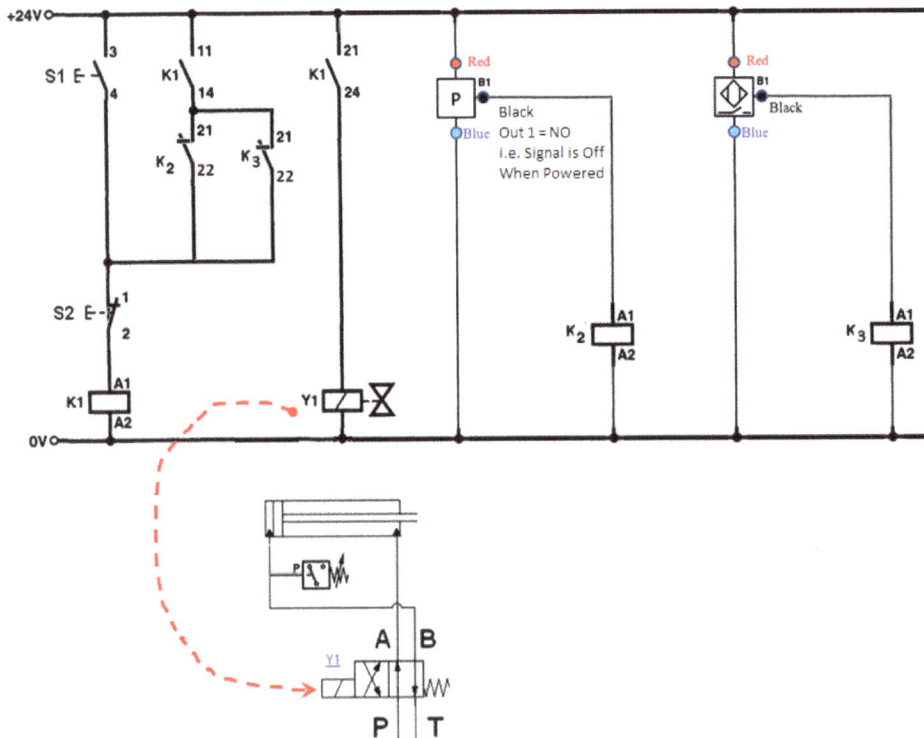

9. A pressure compensated flow control valve isn't function properly in sustaining the motor speed at the desired value. This is likely because of?
   A. Motor is oversized.
   B. The spool of the valve compensator is seized because of contamination.
   C. The pump runs so fast flowing more fluid volume.
   D. None of the above.

10. A proportional pressure relief valve isn't allowing the pressure to build up to desired maximum value. This is likely because of?
    A. The valve isn't receiving control electrical signal.
    B. The valve spool is sized because of the contamination.
    C. An internal spring is broken.
    D. All of the above.

Chapter 7 Assignment

Student Name: -- Student ID: ------------------

Date: --- Score: ------------------------

Question: Describe what could happen if very fine particles built in between the spool and the sleeve of a directional valve,

Chapter 8
Troubleshooting and Failure Analysis of Accumulators

Objectives:

This chapter discusses hydraulic *accumulators* inspection, troubleshooting, and failure analysis. In this chapter, a troubleshooting chart for accumulator faults is presented. The chapter also presents examples of defective accumulators caused by various reasons such as vessel explosion due to material defects and pressure shocks etc.

Brief Contents:

8.1- Hydraulic Accumulators Inspection

8.2- Hydraulic Accumulators Troubleshooting

8.3- Hydraulic Accumulators Failure Analysis

0

0

8.1- Hydraulic Accumulators Inspection

A hydraulic *accumulator:*

- **Function:** is a capacitive device that can store energy →
- **Safety Caution:** review manufacturer's instructions in case of inspection.
- **Configurations:** piston-type, bladder-type, or diaphragm-type.

Volume 5 of these series of textbooks provides information about the construction and operating principles of accumulators.

1

1

Hydraulic Accumulator Inspection Sheet	
Manufacturer	
Model #	
Serial #	
Location	
Accumulator Type	▪ ☐ Piston ☐Bladder ☐Diaphragm
Accumulator Design Values	▪ Nominal Volume: __( ) ▪ Pre-charge Pressure: __( ) ▪ Min System Pressure:__( ) ▪ Max System Pressure: ( )
Current Charge Pressure	▪ Pre-charge Pressure: __( )
Conditions of Bladder/Diaphragm	
Conditions of Accumulator Body	
State of Gas Charge Valve	
Other Notes:	

Table 8.1 – Hydraulic Accumulators Inspection Sheet

2

2

8.2- Hydraulic Accumulators Troubleshooting

T-Accumulator-01: Accumulator Troubleshooting	
Response of accumulator is slow?	▪ Pre-charge gas pressure improperly adjusted? ▪ Piston binding or weak spring. ▪ Relief valve set too low.
Accumulator fails to absorb shocks?	▪ Pre-charge gas pressure is lost or too high. ▪ Diaphragm or bladder ruptured. ▪ Piston seized.
Accumulator isn't properly charged with oil.	▪ Check maximum accumulator's pressure setting. ▪ Check pre-charge gas pressure. ▪ Check leaking accumulator oil discharge valve. **(See Note 1).**
Pre-charge gas pressure improperly adjusted?	▪ Check for possible gas leak from the accumulator. ▪ Check if the bladder or diaphragm is ruptured. ▪ Check the conditions of seals in piston accumulators. **(See Note 2).**

Table 8.2– Accumulator Troubleshooting Chart

3

3

Maximum accumulator's pressure improperly adjusted?	▪ Review maximum pressure and adjust PRV accordingly.
Leaking accumulator oil discharge valve?	▪ Check the if the valve is tightened properly.
Accumulator charging time is unusually long.	▪ Pump undersized. ▪ Pre-charge gas pressure too low. ▪ PRV is set too low or partially stuck open. ▪ Leaking oil-discharge valve or partially opened.
Loose body assembly bolt?	▪ Torque bolts as per the manufacturer's instructions.
Piston binding or seizure in a piston accumulator?	▪ Discharge the gas and oil. ▪ Disassemble and repair.
Bladder/Diaphragm is ruptured?	▪ Check if the compression ratio exceeds the design value **(See Note 3).**

Table 8.2– Continue

4

Note 1 – Accumulator Mounting Manifold: As

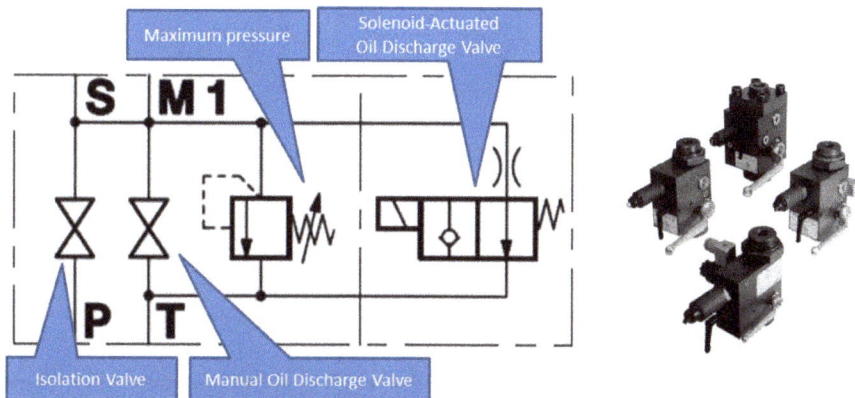

Fig. 8.1 – Mounting Manifold for Accumulators

5

Note 2 – Accumulator Mounting Precharge:

- **Pre-charge pressure too high:**
  - Piston accumulator → the piston travels too close to the end cap.
  - Bladder accumulator → bladder touches the poppet assembly → bladder fatigue failure of the poppet spring assembly, or even a pinched bladder.

- **Pre-charge pressure is too low:**
  - Piston accumulator → the piston is driven into the gas end cap and remains there.
  - Bladder accumulator → the bladder is crushed into the top of the shell → bladder can extrude into the gas stem and be punctured.
  - piston accumulators are generally more tolerant of incorrect pre-charging pressures.

Video 606 (1 min)

6

6

Note 3 – Accumulator Compression Ratio:

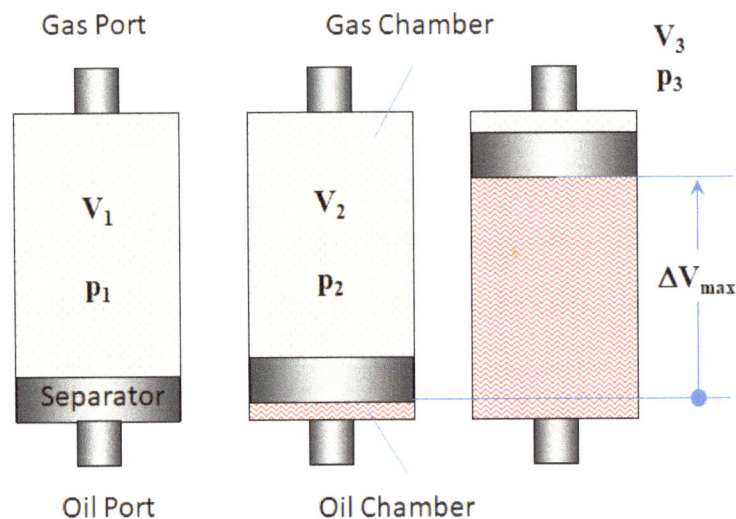

Fig. 8.2 - Operating Principle of Accumulators

7

8.3- Hydraulic Accumulators Failure Analysis

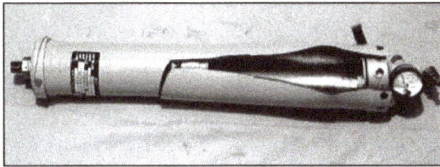

Fig. 8.3 – Accumulator Failure due to Material Defect

Fig. 8.4 – Accumulator Failure due to Exceeding Compression Ratio

Fig. 8.5 – Accumulator Failure due to Accumulator Rupturing

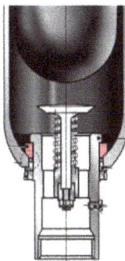

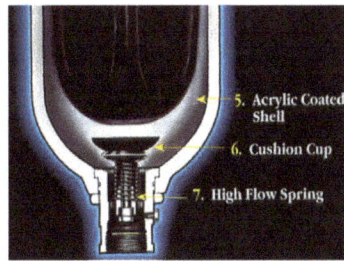

5. Acrylic Coated Shell

6. Cushion Cup

7. High Flow Spring

- A bladder without cushioning cup →
- the bladder wraps around the poppet while fluid is flowing out →
- the poppet closes on the bladder pinching it.

8

8

Chapter 8 Reviews

1. When an accumulator is failed to absorb shocks, this is likely because of?
    A. Pre-charge pressure is set too high.
    B. Pre-charge pressure is set too low.
    C. Diaphragm or bladder are ruptured.
    D. All of the above

2. When an accumulator discharges out the fluid sluggishly, this is likely because of?
    A. Pre-charge pressure is set too high.
    B. Pre-charge pressure is set too low.
    C. Compression ratio exceeds the design value.
    D. All of the above

3. When an accumulator stored less than normal oil volume, this is likely because of?
    A. Pre-charge pressure is set too high.
    B. Maximum system pressure is reduced.
    C. Compression ratio exceeds the design value.
    D. All of the above

4. When a piton accumulator loses gas pressure over the time, this is likely because of?
    A. Diaphragm or bladder are ruptured.
    B. Gas leaks from around the piston.
    C. Gas valve is tightened firmly.
    D. All of the above

5. The shown below accumulator failure is likely because of?
    A. Material defect and overpressure.
    B. Incompatible hydraulic fluids.
    C. Gas valve isn't tightened firmly.
    D. All of the above

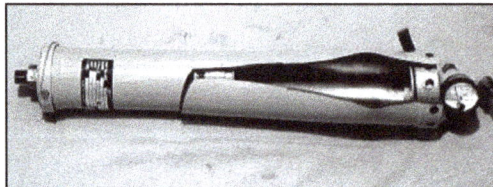

Chapter 8 Assignment

Student Name: --- Student ID: ------------------

Date: -- Score: -----------------------

Question: Describe the safety manifold on which an accumulator should be mounted to ensure safe operation of the accumulator.

Chapter 9
Troubleshooting and Failure Analysis of Reservoirs

Objectives:

This chapter discusses hydraulic *reservoirs* inspection, troubleshooting, and failure analysis. In this chapter, a troubleshooting chart for reservoir faults is presented. The chapter also presents examples of defective reservoirs.

The following topics are discussed in Chapter 2 in Volume 4 "Hydraulic Fluids Conditioning":

- Contribution of Hydraulic Reservoirs.
- Configurations of Hydraulic Reservoirs.
- Construction of Hydraulic Reservoirs.
- Design of Hydraulic Reservoirs.
- Hydraulic Reservoir Design Case Study.

0

0

The following topics are discussed in Chapter 9 in Volume 5 "Maintenance and Safety":

BP-Reservoirs-01-Selection and Replacement.

BP-Reservoirs-02-Maintenance Scheduling.

BP-Reservoirs-03-Installation and Maintenance.

Brief Contents:

9.1- Hydraulic Reservoirs Inspection

9.2- Hydraulic Reservoirs Troubleshooting

9.3- Hydraulic Reservoirs Failure Analysis

1

1

9.1- Hydraulic Reservoirs Inspection

Hydraulic Reservoirs Inspection Sheet	
Manufacturer	
Model #	
Serial #	
Location	
Reservoir Type	• ☐ L-Shaped ☐Foot-Mounted ☐Overhead ☐Compact • ☐ Open to Atmosphere ☐Pressurized
Reservoir Size	• Liters:( ) Gallons:( )
Current Charge Pressure	• Pre-charge Pressure: ( )
Conditions of Breather Filter	
Conditions of Connected Lines	
Conditions of Gaskets	
Conditions of Inside	
Other Notes	

Table 9.1 – Hydraulic Reservoirs Inspection Sheet

2

2

9.2- Hydraulic Reservoirs Troubleshooting

T-Reservoirs-01: Reservoir Troubleshooting	
Leaking reservoir?	• Check gaskets, flanges, other line connections.
Filter breather clogged?	• Replace clogged filter.
Hydraulic fluid aerated?	• Review reservoir design **(See Note 1).** • Consult Chart: • **"T-System-01-Fluid Aeration".**
Hydraulic fluid contaminated by water?	• Check the condition of water absorption cartridge on breather filter. • Check the condition of oil-water heat exchanger. • Cover the reservoir from rain. • Apply appropriate water separation method.

Table 9.2– Reservoir Troubleshooting Chart

3

3

Note 1: Reservoir Design Tips to Help Fluid Deaeration:

- Baffle plate + screen → air removal

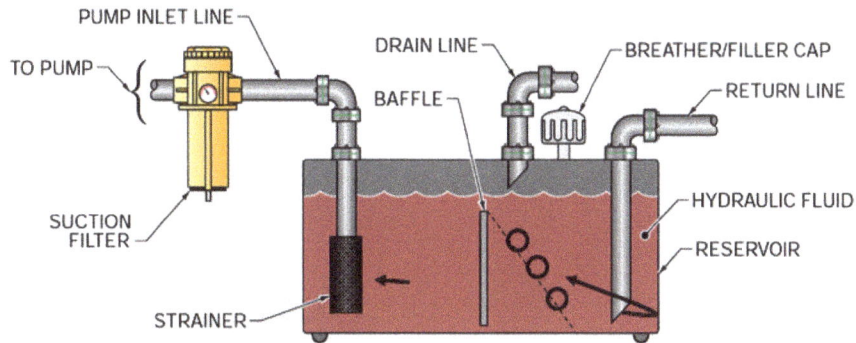

**Fig. 9.1- Role of Hydraulic Reservoir in Fluid Deaeration
(Courtesy of American Technical Publishers)**

4

4

9.3- Hydraulic Reservoirs Failure Analysis

Leaking Gasket:

- Improper gasket material in a hydraulic reservoir of a tractor machine →
- Incompatible with the hydraulic fluid.
- The access hole to the hydraulic reservoir is slowly leaking/dripping.

Fig. 9.2- Leaking Gasket of Hydraulic Reservoir

5

5

Heavily Contaminated Reservoir:

- Hydraulic reservoir is heavily contaminated →
- Rust, stains, and varnish are accumulated on the side walls →
- The overall system required thorough flushing & reservoir cleaning.

Fig. 9.3- Heavily Contaminated Reservoir

6

6

Damaged Attachments and Filling Cap:

- Reservoir is carelessly treated →
- some attachments are bent, and filling cap thread is damaged.

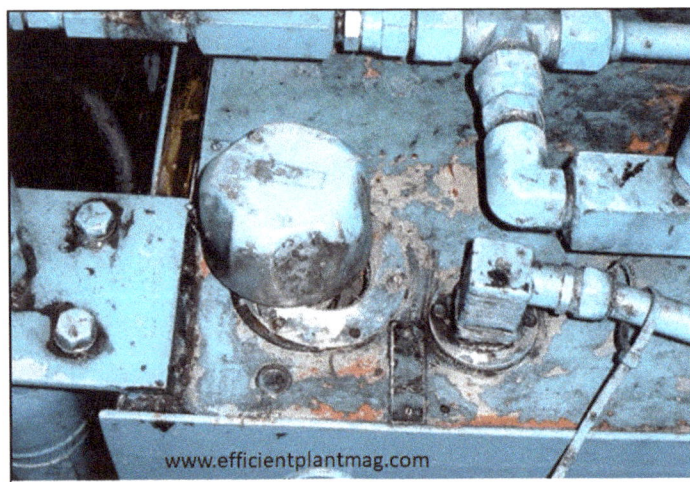

Fig. 9.4- Damaged Attachments and Filling Cap

7

7

Chapter 9 Reviews

1. When the hydraulic fluid temperature increases very fast in the reservoir, this is likely because of?
   A. The reservoir is designed with no baffle plate and both the suction and return pipes are placed closed to each other.
   B. The reservoir designed with regular filling cap with no breather filter.
   C. The reservoir is used in offshore application with no water absorbent air breather.
   D. The return pipe ends on top of the fluid surface and the oil returns back at high speed.

2. When a reservoir the cleanliness level inside the reservoir getting worse over the time?
   A. The reservoir is designed with no baffle plate and both the suction and return pipes are placed closed to each other.
   B. The reservoir designed with regular filling cap with no breather filter.
   C. The reservoir is used in offshore application with no water absorbent air breather.
   D. The return pipe ends on top of the fluid surface and the oil returns back at high speed.

3. When water contamination inside the reservoir is increased more than normal, this is likely because of?
   A. The reservoir is designed with no baffle plate and both the suction and return pipes are placed closed to each other.
   B. The reservoir designed with regular filling cap with no breather filter.
   C. The reservoir is used in offshore application with no water absorbent air breather.
   D. The return pipe ends on top of the fluid surface and the oil returns back at high speed.

4. When fluid is aerated in the reservoir, this is likely because of?
   A. The reservoir is designed with no baffle plate and both the suction and return pipes are placed closed to each other.
   B. The reservoir designed with regular filling cap with no breather filter.
   C. The reservoir is used in offshore application with no water absorbent air breather.
   D. The return pipe ends on top of the fluid surface and the oil returns back at high speed.

5. A clogged strainer inside the reservoir will cause?
   A. Pump cavitation.
   B. Reduce fluid viscosity.
   C. Increase fluid contamination level.
   D. Build water contamination in the reservoir.

Chapter 9 Assignment

Student Name: -- Student ID: ------------------

Date: -- Score: -----------------------

Question: Describe how to design a reservoir to help denigrating the hydraulic fluid.

Chapter 10
Troubleshooting and Failure Analysis of Transmission Lines

Objectives:

This chapter discusses hydraulic *transmission lines* inspection, troubleshooting, and failure analysis. In this chapter, a troubleshooting chart for transmission line faults is presented. The chapter also presents examples of defective transmission lines.

The following topics are discussed in Chapter 3 in Volume 4 "Hydraulic Fluids Conditioning":

- Basic Types and Contribution of Hydraulic Transmission Lines.
- Sizing of Hydraulic Transmission Lines.
- Rated Pressures for Hydraulic Lines.
- Hydraulic Pipes.
- Hydraulic Tubes.
- 3.6- Hydraulic Hoses.
- Flanges for Transmission Line Connections
- Rubber Expansion Fittings.
- Test Points .
- Pressure Measurement Hoses.
- Manifolds.

0

0

The following topics are discussed in Chapter 10 in Volume 5 "Maintenance and Safety"

- 10.1-BP-Transmission Lines-01-Selection and Replacement.
- 10.2-BP-Transmission Lines-02-Maintenance Scheduling.
- 10.3-BP-Transmission Lines-03-Installation and Maintenance.
- 10.4-BP-Transmission Lines-04-Standard Tests and Calibration.
- 10.5-BP-Transmission Lines-05-Transportation and Storage.

Brief Contents:

- 10.1- Hydraulic Transmission Lines Inspection
- 10.2- Hydraulic Transmission Lines Troubleshooting
- 10.3- Hydraulic Transmission Lines Failure Analysis

1

1

10.1- Hydraulic Transmission Lines Inspection

Hydraulic Transmission Lines Inspection Sheet	
Manufacturer	
Model #	
Serial #	
Location	
Transmission Line Type	☐ Pipe ☐Tube ☐Hoses
Transmission Line ID	mm:( ) inches:( )
Transmission Line Length	
Transmission Line Ends	
Conditions of Transmission line	
Conditions of Transmission line	▪ Hose kinks and twists. ▪ Hose cracks from minimum bend radius exceeded. ▪ Hose brittleness or loss of flexibility. ▪ Hose frayed protective layers. ▪ Hose broken reinforcement layers. ▪ Hose outer cover pulled back from coupling ends. ▪ Hose rusted, broken, or loosen ends joints. ▪ Hose abrasion.
Other Notes	

Table 10.1 – Hydraulic Transmission Lines Inspection Sheet

2

2

10.2- Hydraulic Transmission Lines Troubleshooting

Leaking transmission line → cost of:
- Loss of fluid & make-up the fluid.
- Cleaning-up the mess.
- Proper disposal of the spilled fluid in accordance with the local state and federal regulations.
- Contamination ingression.
- Possible safety liability.

Six drops of oil per minute → approximate cost = $1000 per year.

3

T-Transmission Lines-01: Transmission Lines Troubleshooting	
Leaking transmission line? **(See Note 1)**	**▪ Line Body:** ▪ Check if any part of the line body is damaged. ▪ Check if the line is subjected to mechanical stresses. ▪ Check if the line is over pressurized. ▪ Check if the line wasn't cleaned before assembly **(See Note 2).** ▪ Check minimum bend radius for hoses. ▪ Check if a hose is twisted or stretched. ▪ Check if a hose was subjected to abrasion. **▪ Line Fittings:** ▪ Check if line ends are damaged. ▪ Check if line ends are standard and of good quality. ▪ Check tightening torque of line ends. ▪ Flare has cracks or embedded dirt. ▪ Tube is not properly aligned with fitting. ▪ There is no sealant used. ▪ Fitting threads are distorted. ▪ O-Ring leak. **▪ Hydraulic Fluid:** ▪ Verify that the fluid is compatible with the inner tube, the outer cover, fittings, and O-rings.

Table 10.2– Transmission Lines Troubleshooting Chart

4

4

Note 1: Transmission Line Leakage Inspection:
- Leakage inspection by blending special dye with the fluid.
- A specific wavelength lamp → even small leaks become easily visible.
- Various colors for various circuits.

Fig. 10.1- Effective Method for Transmission Line Leakage Inspection (Courtesy of Spectroline)

5

5

Note 2: Hose Leakage due to Line Not Cleaned Before Assembly:

- Improper hose cutting/flushing → metal particles and debris settle inside
- → fractures to develop between fitting and hose assembly →
- hose leakage.

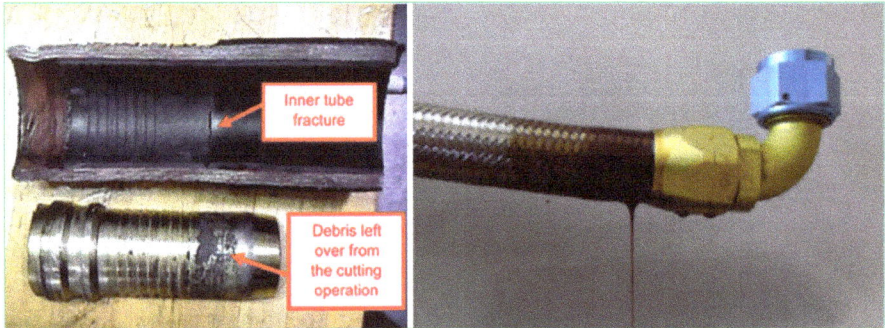

Fig. 10.2– Hose Leakage due to Line Not Cleaned Before Assembly
(Courtesy of Parker)

6

6

10.3- Hydraulic Transmission Lines Failure Analysis

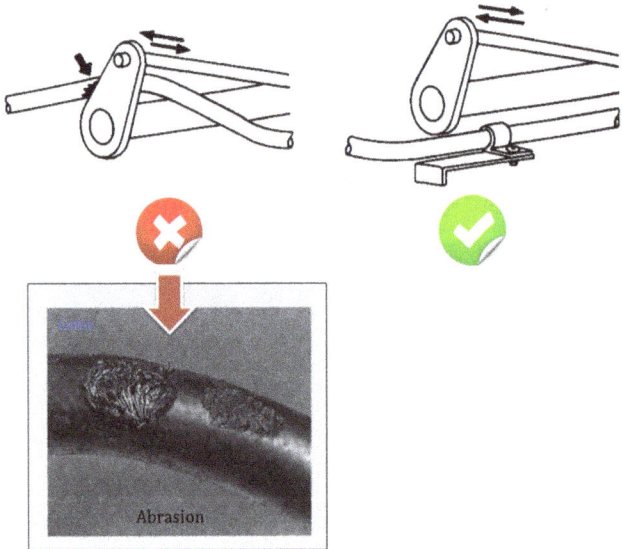

Fig. 10.3– Hose Cover Abrasion (Courtesy of Gates)

7

7

10.3- Hydraulic Transmission Lines Failure Analysis

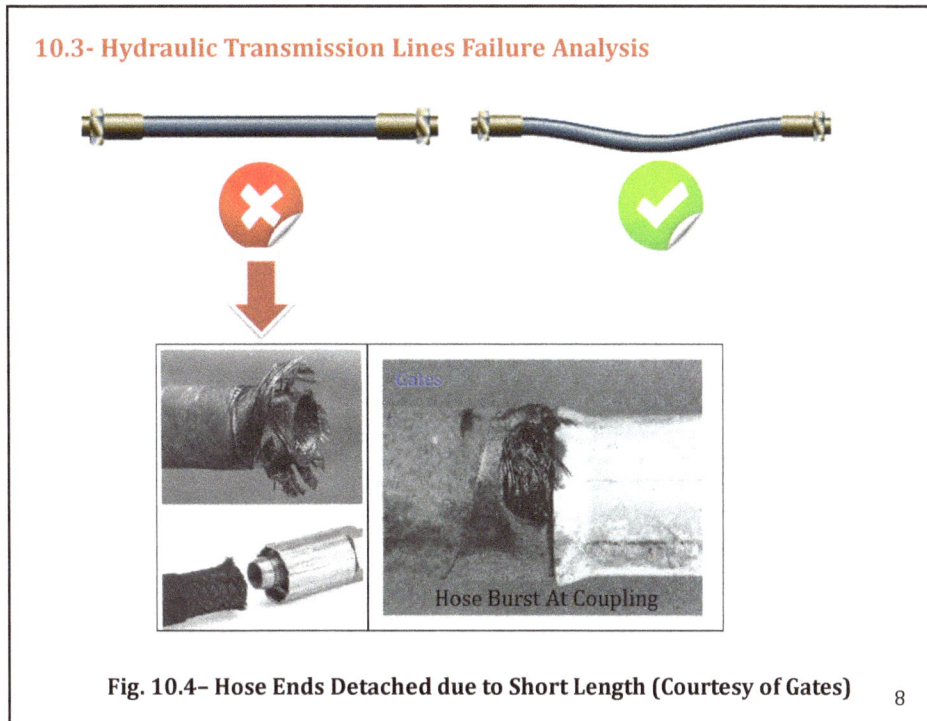

Fig. 10.4– Hose Ends Detached due to Short Length (Courtesy of Gates)

8

8

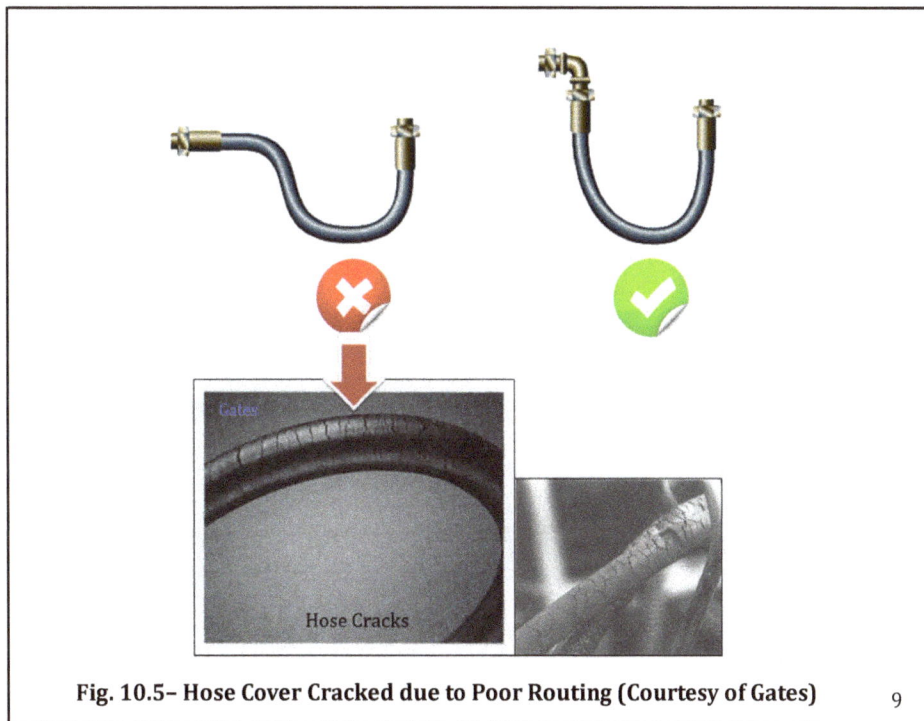

Fig. 10.5– Hose Cover Cracked due to Poor Routing (Courtesy of Gates)

9

9

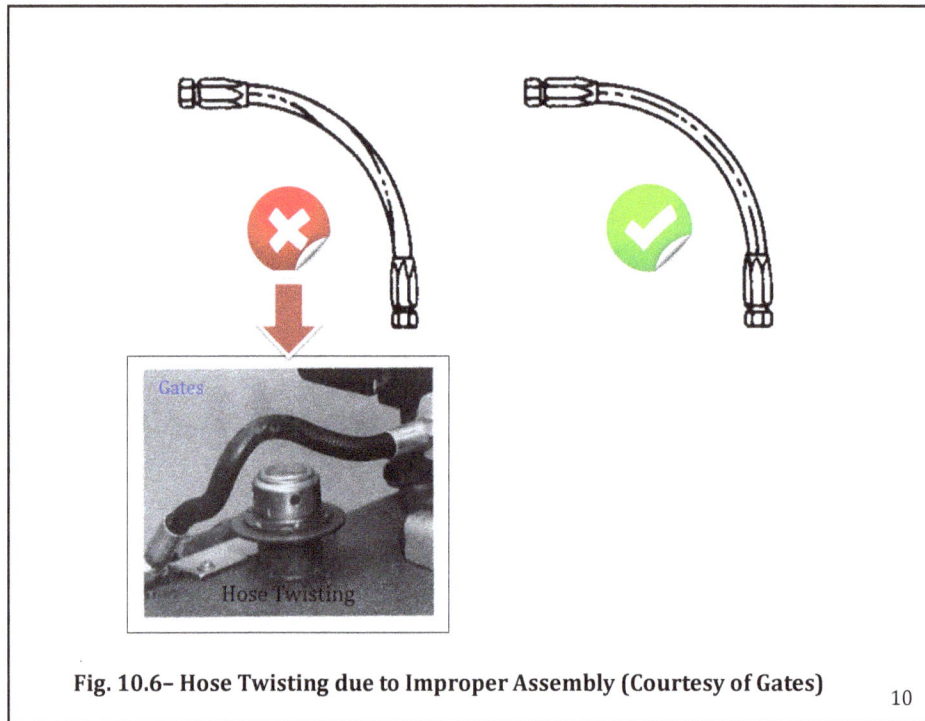

Fig. 10.6– Hose Twisting due to Improper Assembly (Courtesy of Gates)

10

10

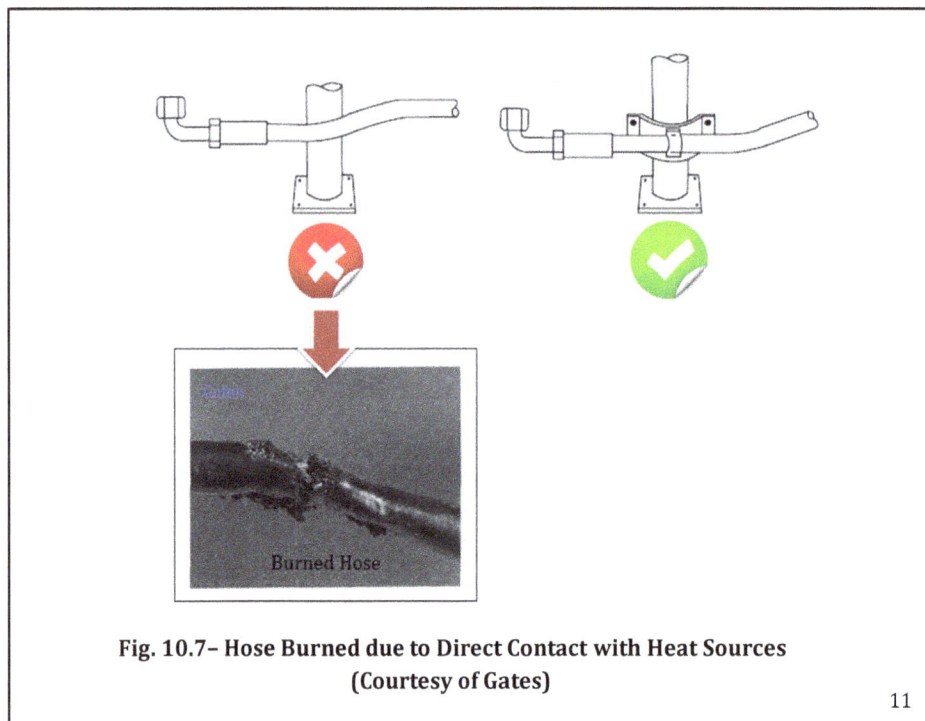

**Fig. 10.7– Hose Burned due to Direct Contact with Heat Sources
(Courtesy of Gates)**

11

11

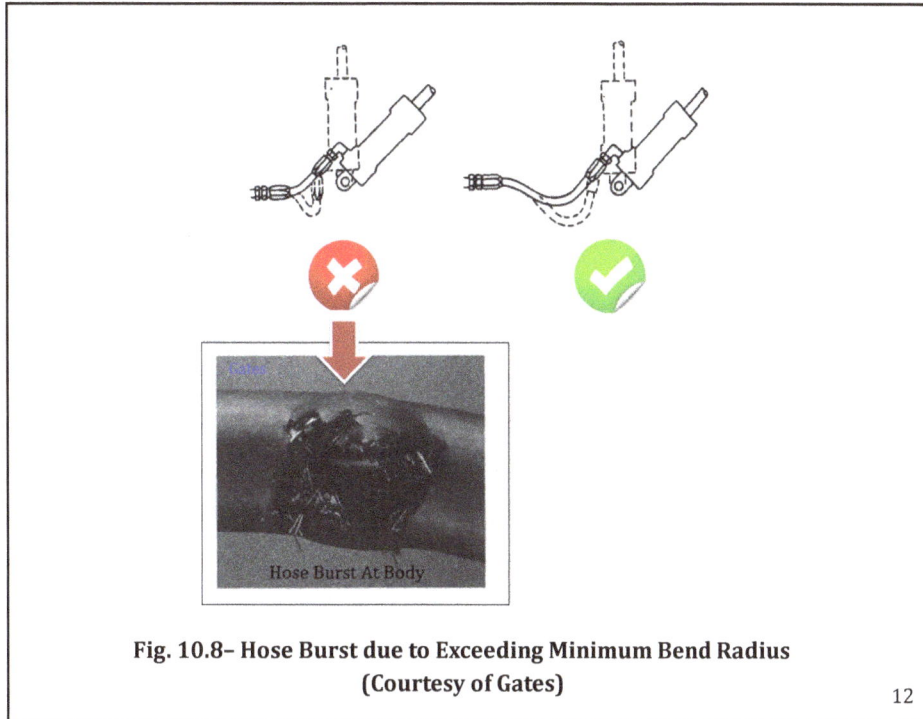

Fig. 10.8– Hose Burst due to Exceeding Minimum Bend Radius
(Courtesy of Gates)

12

12

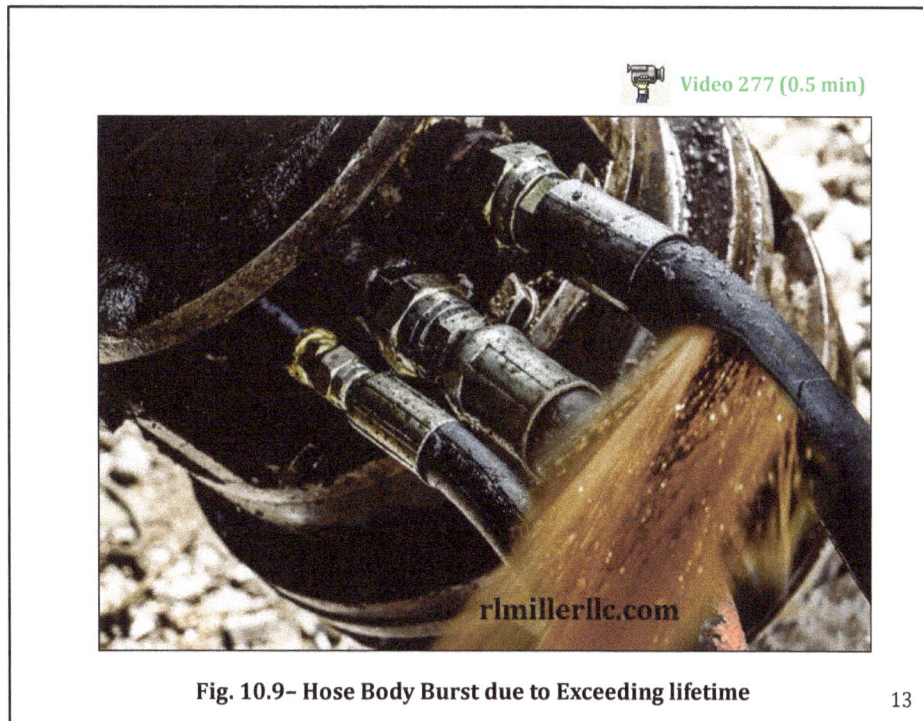

Fig. 10.9– Hose Body Burst due to Exceeding lifetime

13

13

Fig. 10.10– Hose Failure due to Fluid Incompatibility (Courtesy of Parker)

Improper hose fitting assembling →
recommended insertion depth is not met
→ Fitting lose the grip
→ fittings can blow off
→ Dangerous damage could occur.

Fig. 10.11– Hose Ends Blow Off due to Short Length (Courtesy of Parker) 14

14

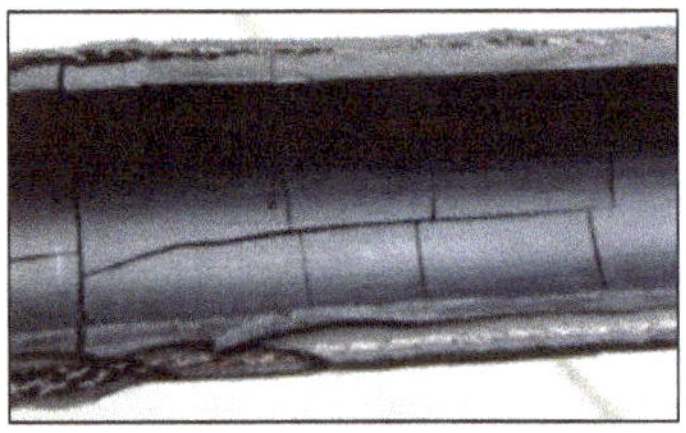

Overheating →
hose to become very stiff →
inner tube is hardened →
Inner tube cracks → hose failure.

Fig. 10.12– Hose Failure due to Overheating (Courtesy of Parker)

Exposure to paint overspray during aircraft manufacturing →
Paint chemically attacks the hose →
Hose is hardened and cracked only on the side that is exposed to paint →
hose lifetime is reduced to half.

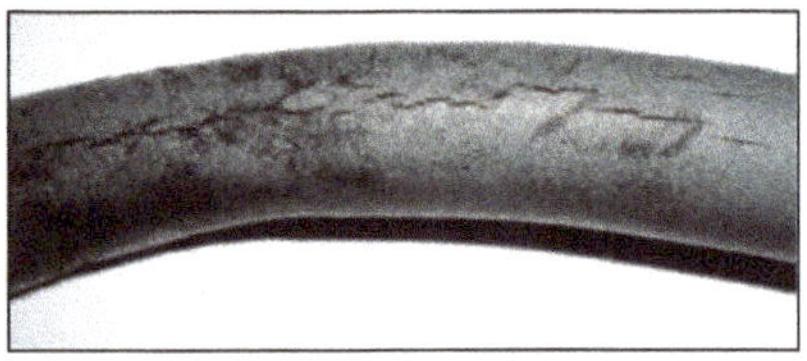

Fig. 10.13– Hose Cracked due to Exposure to Paint Spray 15

15

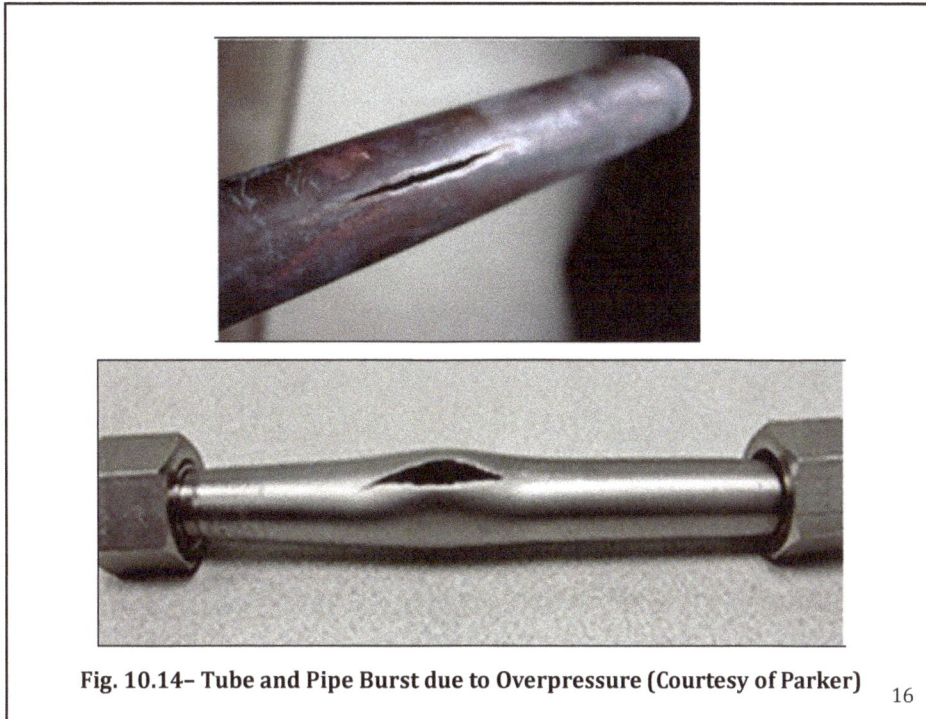

Fig. 10.14– Tube and Pipe Burst due to Overpressure (Courtesy of Parker) 16

16

Fig. 10.15– Pipe and Tube Pin Holes due to Poor Material 17

17

Such pipes must be hanged to the ceiling to release the mechanical stresses.

Fig. 10.16– Pipe and Tube are Leaking due to Mechanical Stresses

18

18

Chapter 10 Reviews

1. The shown below line failure is likely due to?
    A. Abrasion with sharp edges.
    B. Line short length.
    C. Poor routing and exceeding minimum ben radius.
    D. Overpressure.

2. The shown below line failure is likely due to?
    A. Abrasion with sharp edges.
    B. Line short length.
    C. Poor routing and exceeding minimum ben radius.
    D. Overpressure.

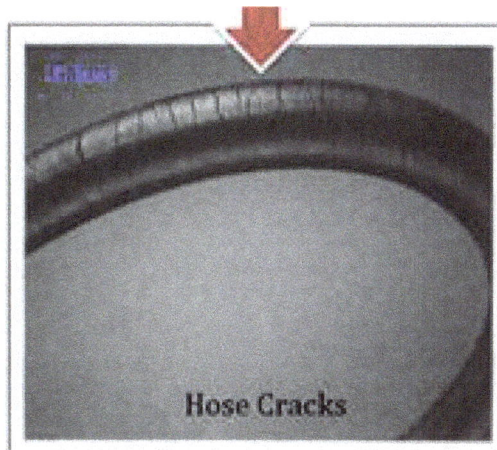

Hose Cracks

3. The shown below line failure is likely due to?
    A. Abrasion with sharp edges.
    B. Line short length.
    C. Poor routing and exceeding minimum ben radius.
    D. Overpressure.

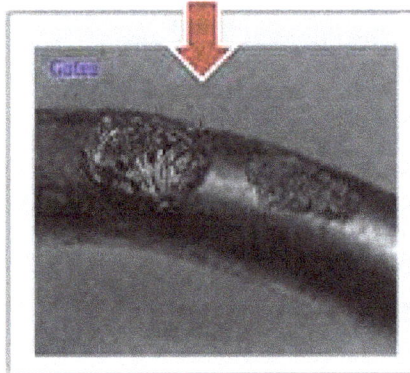

4. The shown below line failure is likely due to?
    A. Abrasion with sharp edges.
    B. Line short length.
    C. Poor routing and exceeding minimum ben radius.
    D. Overpressure.

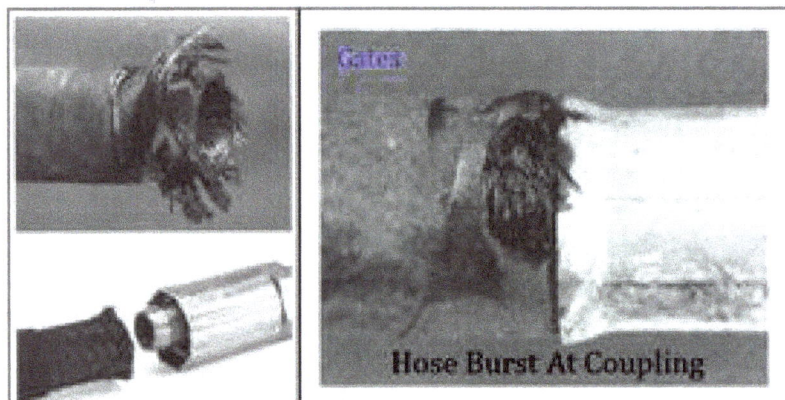

5. A hydraulic tube may leak if?
    A. Subjected to mechanical stresses.
    B. Tube ends are overtightened.
    C. Tube isn't properly aligned with fittings.
    D. All of the above.

Chapter 10 Assignment

Student Name: --- Student ID: ------------------

Date: -- Score: -----------------------

Question: Describe what could happen if line isn't properly cleaned before assembly.

Chapter 11
Troubleshooting and Failure Analysis of Heat Exchangers

Objectives:

This chapter discusses hydraulic *heat exchangers* inspection, troubleshooting, and failure analysis. In this chapter, a troubleshooting chart for heat exchanger faults is presented. This chapter also presents examples of defective heat exchangers.

The following topics are discussed in Chapter 5 in Volume 4
"Hydraulic Fluids Conditioning":

- Contribution of Heat Exchangers.
- Air-Type versus Water-Type Oil Coolers.
- Determination of Cooling Capacity for an Oil Cooler.
- Air-Type Oil Coolers.
- Shell-and-Tube Water-Type Oil Coolers.
- Plat-Type Oil Coolers.
- Cooling-Filtration Units.
- Oil Cooling Circuit Diagram.
- Oil Temperature Automatic Control Solutions.
- Electrical Oil Heaters.

0

0

The following topics are discussed in Chapter 11 in Volume 5
"Maintenance and Safety":

- BP-Heat Exchangers-01-Selection and Replacement.
- BP-Heat Exchangers-02-Maintenance Scheduling.
- BP-Heat Exchangers-03-Installation and Maintenance.
- BP-Heat Exchangers-04-Standard Tests and Calibration.

Brief Contents:

11.1- Heat Exchangers Inspection.

11.2- Heat Exchangers Troubleshooting.

11.3- Heat Exchangers Failure Analysis.

1

1

11.1- Heat Exchangers Inspection

Hydraulic *heat exchangers* can be air-cooled coolers, water-cooled coolers, or heaters. They can work either on/off or proportional to control working temperature within recommended limits for proper operation of hydraulic systems.

Volume 4 and 5 of this series of textbooks presents the construction, operating principles, and guidelines for heat exchangers design, maintenance and safety.

2

2

Hydraulic Heat Exchangers Inspection Sheet	
Manufacturer	
Model #	
Serial #	
Location	
Heat Exchanger Type	☐ Air-Cooled ☐Water-Cooled ☐Heater
Control Mode	☐ On/Off ☐Proportional
Heat Exchanger Cooling/Heating Capacity	
Heat Exchanger Flow	
Conditions of Heat Exchanger	

Table 11.1 – Heat Exchangers Inspection Sheet

3

3

11.2- Heat Exchangers Troubleshooting

T-Heat Exchangers-01: Heat Exchangers Troubleshooting	
Heat Exchanger is among the list of suspicious components that aren't working improperly	▪ Supply Water: ▪ Check adequate water supply flow & temperature. ▪ Check the inlet/outlet temperature of water supply. ▪ Control Parts: ▪ Check adjustment of thermostat. ▪ Check operation of thermostatic water control valve. ▪ Shell/Tube: ▪ Check if the tubes are corroded/eroded. ▪ Air-Cooled Heat Exchangers: ▪ Ambient temperatures too high. ▪ Fan speed or air flow insufficient. ▪ Check if debris has accumulated on tubes or cooling fins.

Table 11.2– Heat Exchangers Troubleshooting Chart

4

4

11.3- Heat Exchangers Failure Analysis

Tube Corrosion:

- Biggest threat oxidation (*corrosion*) →
- Fluid flow is significantly reduced →
- Leaking tubes → Cooling efficiency is reduced.

Fig. 11.1- Corroded Carbon Steel Tube (www.fluiddynamics.com)

Tube Erosion:

- Fluid with (extreme velocities + water contains silica, silt or sea water) →
- Erosion at tube bends (left) and tube ends (right).

Fig. 11.2- Heat Exchanger Tube Erosion (www.fluiddynamics.com)

5

5

Pitting of Tubes:
- Inadequately treated cooling water →
- Presence of chemical compounds (Chloride and Sulphate) →
- Concentrated oxygen (O_2) and carbon dioxide (CO_2) →
- *Chemically-induced* corrosion → tube fails and leaks.

Fig. 11.3- Large Pitting Attack on a Copper Tube (www.fluiddynamics.com)

Thermal Fatigue:
- Thermal stresses + thermal fatigue → tears and cracks

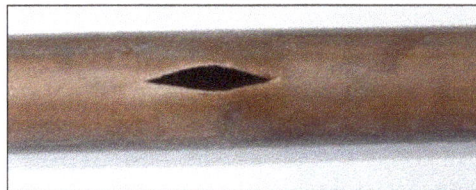

Fig. 11.4- Significant Tear in a Copper Tube due to Extreme Temperature Differences (www.fluiddynamics.com)

6

6

Ice Cold Damage:
- Temperature below freezing → water froze, expanded →
- brass pipes exploded from the inside out.

Fig. 11.5- Ice Cold Damage in Heat Exchangers (Courtesy of Insane Hydraulics)

Steam or Water Hammer:
- Sudden interruption in cooling water flow → pressure surges →
- Water hammer → tubes rupture or collapse.

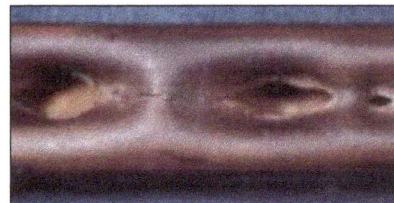

Fig. 11.6- Collapsed and Ruptured Copper Tube Resulting from Steam Hammer (www.fluiddynamics.com)

7

7

Vibration/Resonance:

- Vibration and resonance → tubes rupturing or losing their seal with the tube-sheet and leak.

Fig. 11.7- Broken Baffle and Worn Tubes Caused by Environmental Resonance (www.fluiddynamics.com)

Dusty Air-Cooled Heat Exchanger:

- Accumulated dust on heat exchangers →
- Cooling capacity significantly reduced

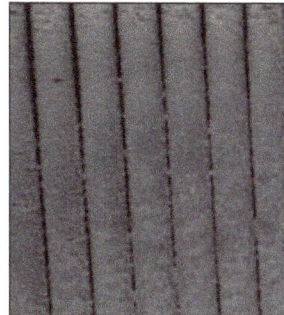

Fig. 11.8- Accumulated Dust on Air-Cooled Heat Exchanger

8

8

Chapter 11 Reviews

1. A water-cooled heat exchanger may function improperly because of?
    A. Change of fluid viscosity.
    B. Change of cooling water inlet temperature.
    C. Change in cooling fan speed and/or insufficient air flow around the radiator.
    D. None of the above.

2. A water-cooled heat exchanger may function improperly because of?
    A. Change of fluid viscosity.
    B. Change of cooling water inlet temperature.
    C. Change in cooling fan speed and/or insufficient air flow around the radiator.
    D. None of the above.

3. The shown below tube failure is due to?
    A. Tube physical erosion due to high fluid flow rate and extreme velocity.
    B. Tube corrosion due to oxidation.
    C. Tube pitting due to chemically induced erosion.
    D. None of the above.

4. The shown below tube failure is due to?
   A. Tube physical erosion due to high fluid flow rate and extreme velocity.
   B. Tube corrosion due to oxidation.
   C. Tube pitting due to chemically induced.
   D. None of the above.

5. The shown below tube failure is due to?
   A. Tube physical erosion due to high fluid flow rate and extreme velocity.
   B. Tube corrosion due to oxidation.
   C. Tube pitting due to chemically induced.
   D. None of the above.

Chapter 11 Assignment

Student Name: --- Student ID: ------------------

Date: -- Score: ------------------------

Question: for an improperly functioning heat exchanger, list what is needed to be checked to detect the fault.

Chapter 12
Troubleshooting and Failure Analysis of Filters

Objectives:

This chapter discusses hydraulic *filters* inspection, troubleshooting, and failure analysis. In this chapter, a troubleshooting chart for filter faults is presented. This chapter also presents examples of defective filter.

Brief Contents:

12.1- Filters Inspection

12.2- Filters Troubleshooting

12.3- Filters Failure Analysis

0

0

12.1- Filters Inspection

Hydraulic *filters* can be suction, pressure, or return. Filters are available in various sizes, efficiency, and dirt holding capacity. Volume 5 of this series of textbooks provides guidelines for filter maintenance and safety.

Filters Inspection Sheet	
Manufacturer	
Model #	
Serial #	
Location	
Filter Type	☐ Suction ☐Pressure ☐Return ☐Breather
Filter Body Type	
Filter Rated Flow Rate	
Filter Beta Ratio/Efficiency	
Conditions of the Filter	

Table 12.1 – Filters Inspection Sheet

1

1

12.2- Filters Troubleshooting

T-Filters-01: Filters Troubleshooting	
• Pressure drop across the filter is larger than the rated value. • OR Clogging Indicator is activated.	• Check if the filter cartridge is clogged due to particulate contaminates, sludge, or varnish. • Check if the flow rate is above the rated value. • Check if the check valve is stuck closed.
• Media cracks. • OR Media migrates downstream the filter.	• Check if filter element is subjected to fatigue due to cyclic flow, such as when a pressure compensated pump is stroked/de-stroked very frequently.
• Improper filtration process.	• Small dirt holding capacity of the cartridge. • Bypass check valve stuck open.
• Broken filter housing.	• Too high pressure. • Shock pressure.

Table 12.2– Filters Troubleshooting Chart

2

2

12.3- Filters Failure Analysis

Filter Clogging due to Particulate Contamination:
- A filter is clogged by dirt (despite it appears clean) →
- By-pass opens due to high differential pressure →
- Unroll the pleated filter media and check what it catches:
  o Metal flacks → machine wear.
  o Rubber flacks → seal deterioration.

Video 286 (1.5 min)

Fig. 12.1- Example of Filter Blockage due to Particulate Contamination (Courtesy of Noria Corporation)

3

3

Filter Clogging due to Sludge:

- High temperatures → fluids break-down →
- Formation of sludge → clogging filters, strainers, and control orifices causing sudden system failure.

Fig. 12.2- Example of Filter Blockage due to Sludge

Filter Clogging due to Varnish:

- High temperatures → chemical degradation → Varnish formation →
- Clogging filters, reduce coolers efficiency, spool seizure, etc.

Fig. 12.3- Example of Filter Blockage due to Varnish

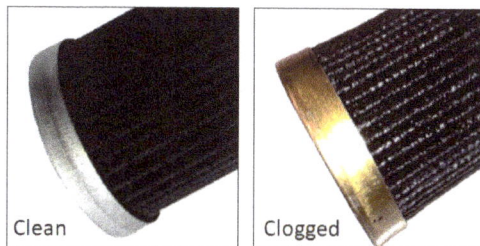

Clean   Clogged

4

4

Filter Media Collapse due to Cyclic Flow:

- cyclic flow (such as when a pressure compensated pump is stroked/de-stroked very frequently) →
- Filter element structural fatigue → pleats cracking

Fig. 12.4- Example of Filter Media Collapse due to Cyclic or Surge Flow

5

5

Chapter 12 Reviews

1. When the pressure drop across a filter increased more than normal, this is likely because?
    A. Filter cartridge is clogged.
    B. Filter is undersized below the design sized.
    C. Oil viscosity is increased above design value.
    D. All of the above.

2. The shown below filter failure is likely because of?
    A. Sludge.
    B. Varnish.
    C. Particulate contaminants.
    D. Flow fatigue.

3. The shown below filter failure is likely because of?
    A. Sludge.
    B. Varnish.
    C. Particulate contaminants.
    D. Flow fatigue.

Clean   Clogged

4. The shown below filter failure is likely because of?
    A. Sludge.
    B. Varnish.
    C. Particulate contaminants.
    D. Flow fatigue.

5. The shown below filter failure is likely because of?
    A. Sludge.
    B. Varnish.
    C. Particulate contaminants.
    D. Flow fatigue.

Chapter 12 Assignment

Student Name: --- Student ID: -------------------

Date: -- Score: ------------------------

Question: Explain reasons of sludge formation around a filter element.

Chapter 13
Hydraulic Systems Troubleshooting

Objectives:

This chapter introduces troubleshooting charts for failures of generic hydraulic systems. Each troubleshooting chart includes relevant notes and examples for better understanding.

0

Brief Contents:

13.1-Features of Hydraulic Systems Failures
13.2-Main Causes of Hydraulic Systems Failures
13.3- Fluid Aeration
13.4- Pump Cavitation
13.5- Excessive System Noise & Vibration
13.6- Excessive System Heat
13.7- Low Power System
13.8- Faulty System Sequence
13.9- External Leakage
13.10- Troubleshooting of Open Hydraulic Circuits
13.11- Troubleshooting of Closed Hydraulic Circuits (Hydrostatic Transmissions)
13.12- Actuator Slow Performance
13.13- Actuator Fast Performance
13.14- Actuator Erratic Performance
13.15- Actuator Moves in Wrong Direction
13.16- Actuator Stops to Move
13.17- Actuator Load Drifts
13.18- Actuator Leaks

1

13.1-Features of Hydraulic Systems Failures

- **Closed System:** Can't see inside the system.
- **Common Symptoms:** Same symptoms could be due to different causes.
- **Failures are Transferrable:** problem transferred with the fluid.
- **Failures are Accelerated:** The rate of growth isn't linear.
- **Chain Action: e.**g. (internal leakage – heat – poor viscosity – less lubrication – wear – internal leakage).

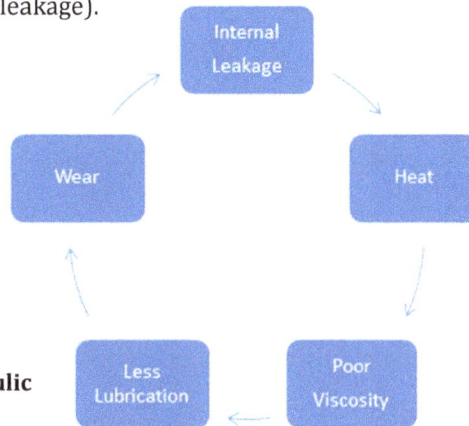

Fig. 13.1- Chain Action of Hydraulic System Failures

2

2

13.2-Main Causes of Hydraulic Systems Failures

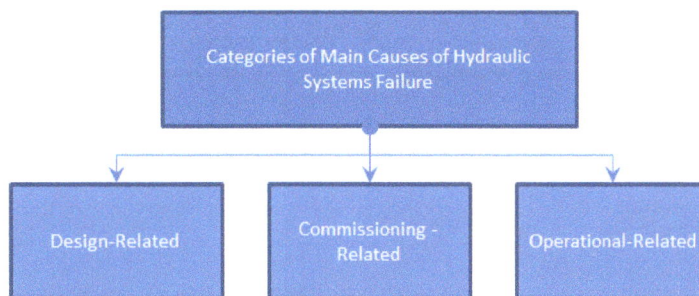

Fig. 13.2- Categories of Main Causes of Hydraulic System Failures

13.2.1-Design-Related Failure Causes

- Incorrect Component Sizing.
- Lack of Predicting System Performance.
- Specifying Wrong Fluid:
- Poor Filtration System Locations.
- Poor Reservoir Design.

3

3

13.2.2-Commissioning-Related Failure Causes

- Well designed system can still fail if it isn't commissioned properly
- Improper commissioning → failures may arise →
- For example:
  - Improper routing of hydraulic lines → lots of line losses.
  - Improper assembly of vanes in a vane pump → in loss of flow.
  - Improper assembly of a pump → pump cavitation.
  - Shaft misalignment in pumps and motors → vibration and shaft seal failure.
  - No or improper pump priming → premature failure.

- Therefore, review the assembly instructions provided by the manufacturer.

4

4

13.2.3-Operationl-Related Failure Causes

- Well designed and commissioned system can still fail if it isn't operated properly

- **For example:**
  - Lack of Understanding.
  - Lack of Maintenance.
  - Low Fluid Volume:
  - Changing Fluid Timely:
  - Keeping working temperature within allowable limits (low – high)
  - Fluid Contamination.
  - Misuse (overspeed & overpressure).

5

5

13.3- Fluid Aeration

Air Leaks into System:

- Insufficient oil volume + leak in suction side → Air leak into system.

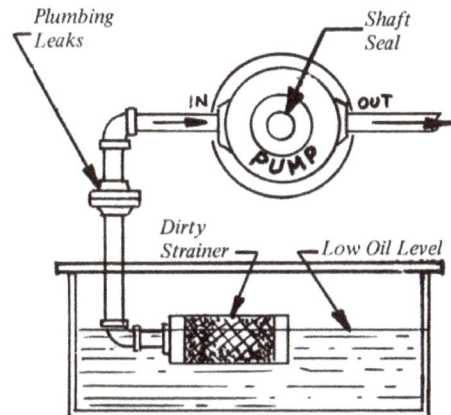

**Fig. 13.3- Sources of Fluid Gaseous Contamination
(Courtesy of Womack)**

6

6

Air Separation from Fluid:
- Fluids contain (7-10)% by volume homogeneously dissolved air.
- Negative pressure zone → dissolved air separates in form of bubbles.

Hydraulic Fluid Evaporation:
- Severe temperature and vacuum conditions → fluid evaporates →
- cloudy to creamy color fluid appearance.

Aeration:
- Sustained tiny emulsified bubbles below the surface of the fluid →
- Milky appearance (oil will become clear in about an hour after shut-down.

Foaming: Accumulation of bubbles on top of the fluid surface.

7

7

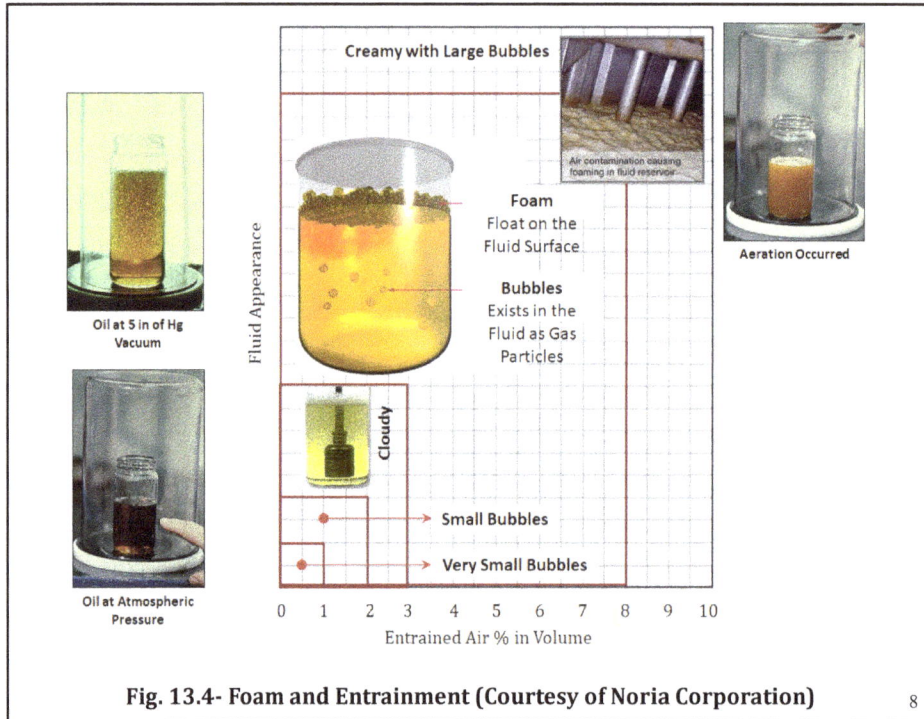

Fig. 13.4- Foam and Entrainment (Courtesy of Noria Corporation)

8

Result of Aeration: 🔊 Sound 02: Aeration

- Power Loss.
- Noisy and vibrated operation.
- Poor lubrication.
- Increased rate of oxidation.
- Erratic/Sluggish machine motion and control.
- Drastic reduction in fluid's bulk Modulus.

Viscosity	Release Time
ISO VG 10, 22, and 32	Maximum 5 min
ISO VG 46 and 68	Maximum 10 min
ISO VG 100	Maximum 14 min

**Table 13.1- Air Separation Capacity in Minutes at 50 °C
(Courtesy of Bosch Rexroth)**

9

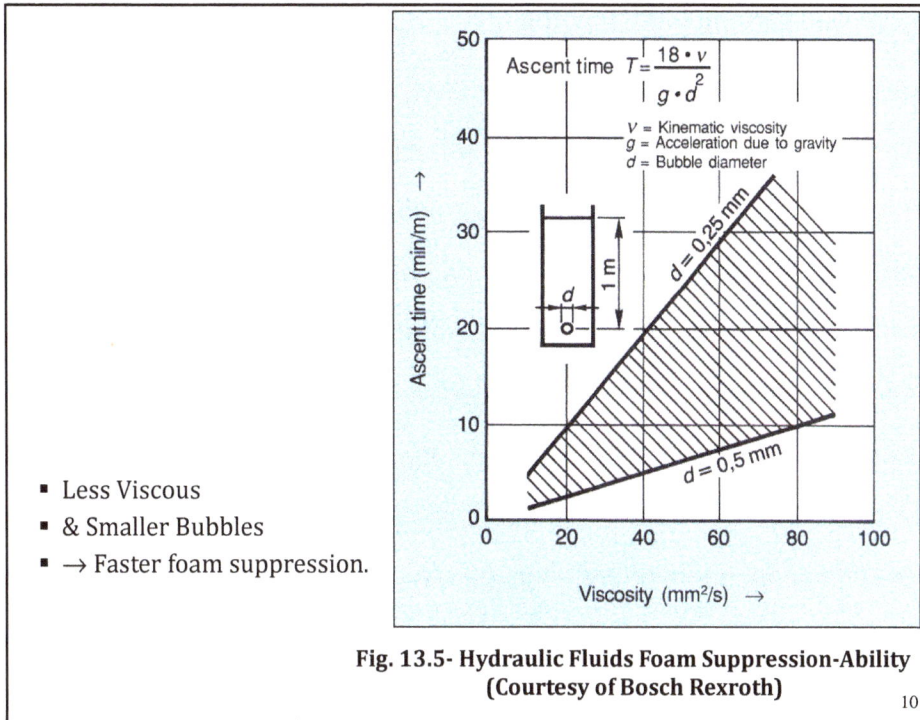

- Less Viscous
- & Smaller Bubbles
- → Faster foam suppression.

Fig. 13.5- Hydraulic Fluids Foam Suppression-Ability
(Courtesy of Bosch Rexroth)

10

10

T-System-01-Fluid Aeration	
Reservoir	
Reservoir fluid level too low?	▪ Follow the guidelines to make up the oil in the reservoir to the specified level.
Poor reservoir design? Return lines were improperly located?	▪ Review reservoir design about air removal. ▪ Consider using screen separator placed on 45-degree angle between suction and return side. ▪ Consider using baffle plate between suction and return lines to elongate oil residence time in the reservoir. ▪ Consider return line discharges below the oil surface to reduce fluid vertexing and sloshing.
Transmission Lines	
Fluid returns at high speed?	▪ Resize the return line so that the return fluid speed is (1.2 - 2.1) m/s = (4 - 7) ft/s. Using return diffuser helps getting rid of air.

Table 13.2- Troubleshooting Chart (T-System-01-Fluid Aeration)

11

Turbulent flow in the system?	▪ Review system design to secure laminar flow (flow rate – conductor size – fluid viscosity).
Air Leaking	
Air Leaks into the Pump? 📹 Video 169 (0.5 min)	▪ Consult Chart: ▪ **"T-Pump-11-Air Leaks into the Pump"**
Air Leaks into the cylinder?	▪ Check if air leaking into cylinder through a rod seal when an overrunning load drives the cylinder developing vacuum.
Gas leaks from the accumulator?	▪ Consult Chart: ▪ **"T-Accumulator-01-Accumulator Troubleshooting"**
Hydraulic Fluids	
Foam suppression additives are depleted.	▪ Consult the fluid supplier about the best practices for adding foam suppressors to the fluid.

Table 13.2- Continue

12

12

13.4- Pump Cavitation

Cavitation Mechanism:
- Is always been confused with aeration.
- Liquid pressure (at a given temperature) = vapor pressure of the fluid →
- Formation of bubbles ate suction side) →
- Collapse of bubbles in pressure side.

📹 Animation-005

Fig. 13.6- Pump Destruction due to Cavitation (Courtesy of Assofluid)

13

13

Subsequently:

- super-compressed (imploded) with microjet action →

Video 208 (1.5 min)

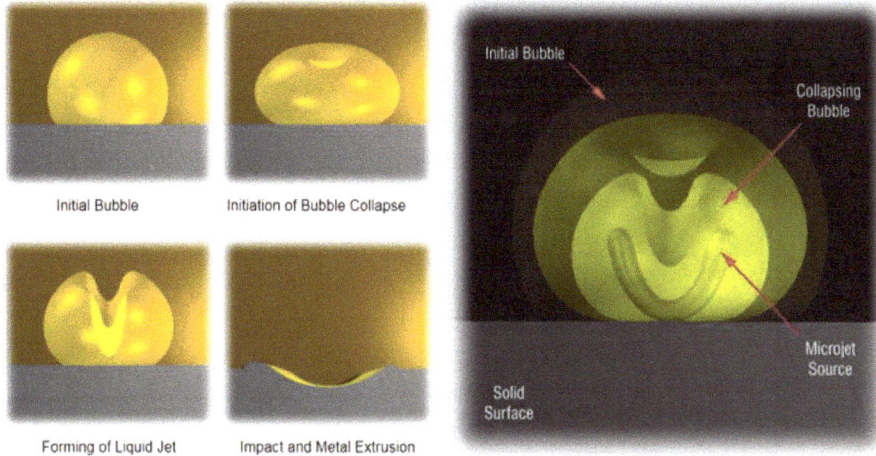

Fig. 13.7- Pump Cavitation (Courtesy of Noria)

14

14

Results of Cavitation:

Video 281 (4 min)

- Same consequences of aeration +
- Implosion of bubbles → microjet shock load →
- Unrepairable material destruction (see below cavitation resistance)
- Sound emission + pump vibration + Loss of pump flow and pressure.
- Dieseling effect on seals leaving burning spots.

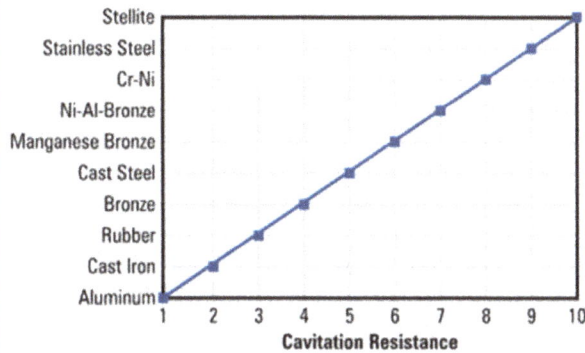

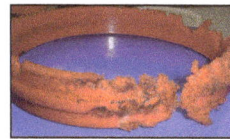

Fig. 13.8- Cavitation Resistance (Courtesy of Noria)

15

15

Cavitation Avoidance:

❖ <u>System Design Considerations:</u>
- Oil Reservoir Design → remove bubbles and reduce turbulence.
- Boosting Pump Intake → protect large size and high-speed pumps.
- Vacuum Switch → preventive solution to stop the pump if cavitated.

❖ <u>Pump Installation Considerations:</u>
- Pump Placement: below fluid surface to build positive intake pressure.
- Intake Line: Properly sized & routed
- Pump Priming: review manufacturer instructions for proper priming.
- Suction Strainer: review manufacturer instructions for sizing.

❖ <u>System Operation Considerations:</u>
- Working Temperature: keep within recommended range.
- Hydraulic Fluid: recommended viscosity.
- Driving Speed: keep it within recommended range.
- Suction Strainer/Suction Line: inspect and clean periodically.
- Oil Level: inspect periodically.

Video 168 (2 min)

16

16

T-System-02-Pump Cavitation	
Reservoir: Fluid level is too low, reservoir with no baffle, or too shallow?	▪ Follow the guidelines for reservoir design and make up the oil to the specified maximum level.
Reservoir isn't vented?	▪ Check clogged air breather in open tanks. ▪ Check pressure of closed reservoir.
Suction Line: Suction valve is partially or fully closed?	▪ Fully open the suction valve. ▪ Lock the suction valve in fully opened position.
Suction filter/strainer/air breather are clogged or undersized?	▪ Wash or replace strainer or suction filter. ▪ Check manufacturer recommendation about using suction strainer for this pump.
Suction line is undersized, kinked, restricted, sucking air, or plugged.	▪ Check intake line size vs. pump intake port. ▪ Flow speed = (0.6 - 1.2) m/s = (2 - 4) ft/s.

Table 13.3- Troubleshooting Chart (T-System-02-Pump Cavitation)

17

Suction line is too long, too short, or has too many bends?	▪ Review design of intake line (Chapter 2 – Volume 4)
Is the suction line flexible hose?	▪ Make sure it is a suction hose (not a pressure hose) because inner layers of pressure hoses are not made to carry negative pressure.
Hydraulic Fluid: Fluid is too hot?	Consult Chart: **"T-System-04-Excessive System Heat"**
Fluid is too cold or too viscous?	▪ Review the pour point of the fluid. ▪ Warm the fluid 10 degrees C above the pour point before starting the machine. ▪ Review manufacturers recommendation about the fluid viscosity and replace the fluid if needed.
Pump: Pump is placed too high from fluid surface or in a wrong orientation?	▪ Review the manufacturer's instructions about the pump maximum suction head or pump orientation (review Volume 5).
Does the pump rotate at high speed?	▪ Review the manufacturer's instructions about the maximum allowable speed.
Is the pump supercharged?	Check the pump inlet pressure and the inspect the boosting pump.

Table 13.3- Continue

18

18

13.5- Excessive System Noise & Vibration

Vibration → gradually loosens the fittings → fatigue failure of lines.

Noise (structure-borne, fluid-borne and air-borne) → Environmental pollution + long-term personal disability.

OSHA Standard (Act of 1970).

Hours/Day	Sound Level (db.)
8	90
6	92
4	95
3	97
2	100
1-1.5	102
1	105
0.5	110
0.25 or less	115

Table 13.4– Permissible Noise Exposure

19

19

T-System-03-Excessive System Noise & Vibration	
Hydraulic Fluid:	
Fluid Aeration (Is there air in the system or fluid looks milky)?	• Consult Chart: • "T-System-01-Fluid Aeration".
Machine is too cold, or fluid viscosity is too high?	• Check fluid viscosity and resolve accordingly. • Check recommended temperature and resolve accordingly.
Transmission Lines:	
Is the pipework adequately supported?	• Support the pipework as per the installation guidelines.
Pump:	
Pump is noticeably noisy and vibrating?	• Consult Charts: • "T-Pump-12-Excessive Pump Noise and Vibration".
Other Units:	
Electrohydraulic valve shuddering?	• Defective solenoid or voltage too low. • Large flow and flow forces through the valve. • Improper control settings.
Is the noise associated to a specific unit?	• Consult Chart • "T-Unit-02-Noisy Unit".

Table 13.5- Troubleshooting Chart
(T-System-03-Excessive System Noise & Vibration)

20

20

13.6- Excessive System Heat

Sources of Heat in Hydraulic Systems:
- Broadly classified as (Design-Related and Operation-Related).
- Approximately 25 percent of the input power is converted to heat.
- Operator bad habits → system overheating.

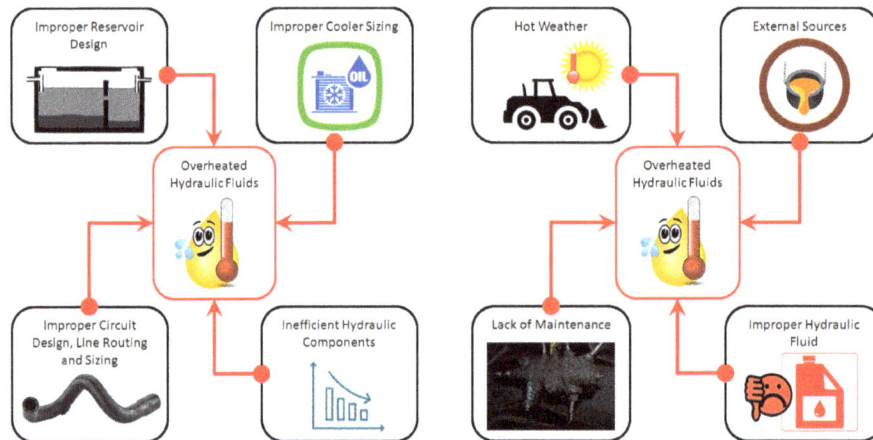

Fig. 13.9- Design-Related Heat Sources   **Fig. 13.10- Operation-Related Heat Sources**

21

21

Consequences of Hydraulic System Overheating:

- System Overheating → reduced oil viscosity →
  - Reduced lubrication → metal-to-metal friction.
  - Increased internal leakage → inefficient system.
- System Overheating → fluid degradation → accelerated oxidation sludge → sludge formation.

❖ **Petroleum-Based Fluids:**
- Optimum Temperature: typically, from 38°C to 54°C (100°F to 130°F).
- Critical Temperature: 70°C (158°F).
- Every 10°C (18°F) > critical → oxidation ↑ (doubled) & oil life is ↓ (half)
- Working consistently at 80°C (176°F) → fluid life ↓ (75%).
- Working above 82°C (180°F) → damages most seal compounds.

❖ **Water-Based Fluids:**
- Overheating → water evaporation → ratio of [water/(base fluid)] ↑ → viscosity and additive concentration ↑ fluids fire resistance↓.

22

22

T-System-04-Excessive System Heat	
Oil Cooler: Insufficient cooling capacity?	Check: ▪ Cooling water supply temperature and valve. ▪ Thermostat adjustment/operation. ▪ Cleanliness of the cooler (air/water). ▪ Dirt accumulated on top of outer surfaces of reservoir, lines, and other components.
Reservoir: Reservoir fluid level too low?	▪ Follow the guidelines to make up the oil in the reservoir to the specified level.
Poor reservoir placement?	▪ Avoid placing reservoir in a point of less air flow. ▪ Use forced air ventilation (e.g. industrial fan) to drive airflow around the reservoir. ▪ Temperature at the exterior of the reservoir should not exceed 140ºF (60ºC). ▪ Exterior of reservoir and all components must be kept clean to ensure that no hot spots develop as a result of accumulated dust and dirt.

Table 13.6- Troubleshooting Chart (T-System-04-Excessive System Heat) 23

23

Poor reservoir design?	▪ Review guidelines of placing suction line and return line. ▪ Use of baffle plate to separate suction line from return line. ▪ Reservoir size it too small.
Hydraulic Fluid: Improper fluid conditions?	▪ Check fluid viscosity, cleanliness, and aeration.
Working Pressure: System pressure above normal?	▪ Check if load on actuators above normal. ▪ Reset maximum system pressure if possible.
Working Flow: Excessive flow over relief valve?	▪ Check: relief valve setting, pump driving speed, maximum flow of variable displacement pump, and system duty cycle. ▪ Pump may not be unloaded between cycles. ▪ Compensator of pressure-compensated pump may be set higher than the relief valve.

Table 13.6- continue

24

24

Transmission Lines: Undersized transmission lines?	▪ Undersized transmission lines cause turbulent flow and increase wasted energy and heat generation.
Flow Control Valves:	▪ Check if the valve is undersized. ▪ Check if needle valve is installed backward.
Directional Control Valves:	▪ Check if the valve is undersized.
Pressure Relief Valve	▪ Check if the valve is misadjusted, so it is leaking oil to the tank.
Pressure Compensated Pump Pressure compensator is misadjusted.	▪ Check and readjust.
Other Units: Any sign of worn component and/or internal leakage?	▪ Examine and test valves, cylinders, motors, etc. for internal leaks. ▪ If wear is abnormal, replace the component at fault.
Heat is associated with a specific component?	▪ Consult Chart: ▪ **"T-Unit-03-Excessively Hot Unit".**

Table 13.6- continue

25

25

13.7- Low Power System

(A machine is barely carrying a load + actuator moves when the load is reduced) sign for reduced load carrying capacity.

Figure 13.11- Low Power System

26

26

T-System-05-Low Power System	
Engine & Power Transmission: Insufficient Power and Torque at the Power Take-offs?	▪ Air filter of the engine is clogged. ▪ Power transmission defective (e.g. V-belt or toothed belt slippage, key sheared off at pump-motor coupling). ▪ Check reference duty cycle of the machine.
Pump: Pump outlet low pressure?	▪ Consult Chart: ▪ **"T-Pump-06-Low Pressure at the Pump Outlet".**
Actuator: Actuator Leaks?	▪ Consult Chart: ▪ **"T-Cylinder-1-Cylinder Troubleshooting".**
DCV: DCV at Fault?	▪ Consult Chart: ▪ **"T-Valve-01-DCV Troubleshooting".**
PRV: PRV at Fault?	▪ Consult Chart: ▪ **"T-Valve-03-PRV Troubleshooting".**
Filter: Pressure filter blocked?	▪ Check and act accordingly.

Table 13.7- Troubleshooting Chart (T-System-05-Low Power System)

27

27

13.8- Faulty System Sequence

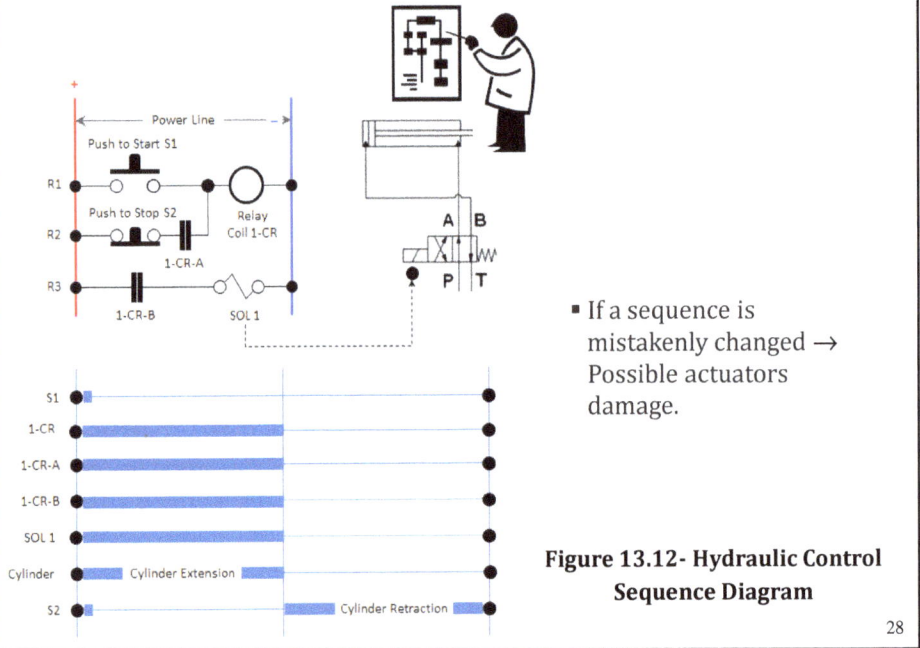

- If a sequence is mistakenly changed → Possible actuators damage.

Figure 13.12- Hydraulic Control Sequence Diagram

28

28

T-System-06-Faulty System Sequence	
Actuator: Actuator Slow Performance?	▪ Consult Chart: ▪ **"T-System-10-Actuator Slow Performance".**
Actuator Fast Performance?	▪ Consult Chart: ▪ **"T-System-11-Actuator Fast Performance".**
Actuator Erratic Performance?	▪ Consult Chart: ▪ **"T-System-12-Actuator Erratic Performance".**
Actuator Moves in Wrong Direction?	▪ Consult Chart: ▪ **"T-System-13-Actuator Moves in Wrong Direction".**
Actuator Stops to Move?	▪ Consult Chart: ▪ **"T-System-14-Actuator Stops to Move".**
DCV: DCV at fault?	▪ Consult Chart: ▪ **"T-Valve-01-DCV Troubleshooting".**
Controller: System controlled by a PLC or host controller?	▪ Check control code. ▪ Check digital/analog signals out of the controller.
System contains sensors	▪ Check wiring of sensors. ▪ Check proper functioning of sensors.

Table 13.8- Troubleshooting Chart (T-System-06-Faulty System Sequence)

29

29

13.9- External Leakage

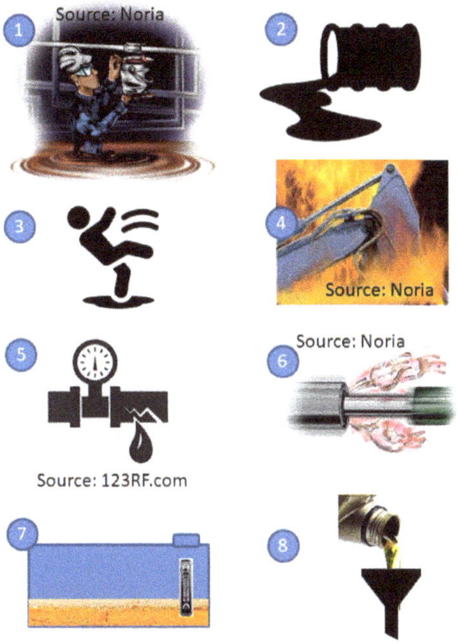

1. Cost of making up the lost fluid.

2. Cost of removing environmental pollution.

3. Risk of personal safety due to slipping.

4. Risk of personal safety due to fire hazard.

5. Loss of system pressure.

6. Loss of actuator power.

7. Loss of oil volume → temperature↑.

8. New fluid → contamination↑.

Figure 13.13- Examples of External Leakage

30

30

T-System-07-External Leakage	
Working Pressure: System pressure is too high?	▪ Review working pressure and resolve problem accordingly.
Working Temperature: Excessively hot system?	▪ Check the deterioration of elastomeric seals because of elevated fluid temperatures.
Transmission Lines: Leakage is traced to a certain fitting or a conductor?	▪ Check if a fitting is loose or over tightened. ▪ Check if different brands of fittings are mixed. ▪ Check if a cutting ring or sealing element is reused. ▪ Check if a conductor is damaged. ▪ Check if a conductor is improperly clamped. ▪ Check if a conductor is improper installed leaving mechanical stress in it. ▪ Check if a tube is properly flared.
Cylinder: Leaking Cylinder?	▪ Consult Chart: ▪ **"T-System-15-Actuator Load Drifting".**
Hydraulic Fluids: Incompatible fluid?	▪ Check the compatibility of the fluid with the shaft seals of pumps, motors, and cylinders.
Other Units: Leakage is traced to one unit.	Check the seal condition of the leaking component.

Table 13.9- Troubleshooting Chart (T-System-07-External Leakage) 31

31

13.10- Troubleshooting of Open Hydraulic Circuits

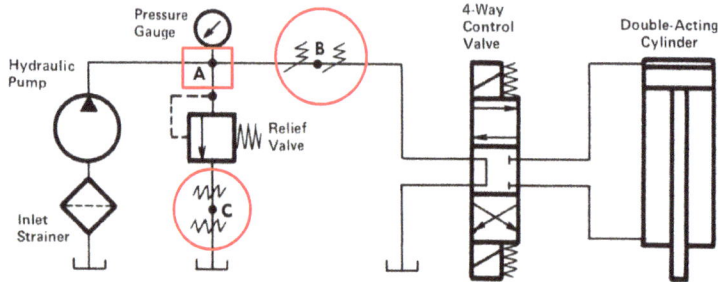

**Figure 13.14- Typical Circuit Diagram of Open Hydraulic Circuit
(Courtesy of Womack)**

T-System-08-Troubleshooting of Open Hydraulic Circuits	
Improper operation of an open circuit	▪ Step 1: Check Pump Inlet Strainer ▪ Step 2: Check Pump and Relief Valve: ▪ Step 3: Check Pump: ▪ Step 4: Check Pressure Relief Valve: ▪ Step 5: Check Cylinder: ▪ Step 6: Check Directional Control Valve:

**Table 13.10- Troubleshooting Chart
(T-System-08-Troubleshooting of Open Hydraulic Circuits)** 32

32

❖ **Step 1: Pump Inlet Strainer:**

▪ **Fault:** A dirty strainer → pump cavitation and permanent failure.

▪ **Location:**

o Commonly It is located inside the reservoir under the oil level.

o Sometimes it is located outside the reservoir on the intake line.

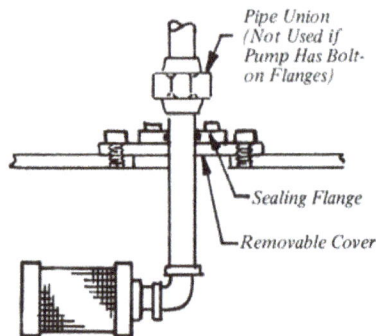

Uncouple inlet line, remove cover plate, and withdraw strainer from reservoir.

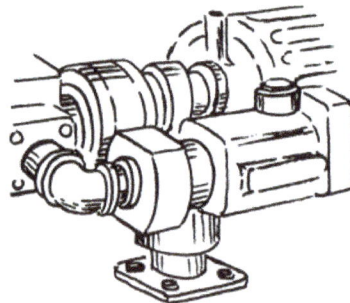

On externally mounted pump inlet strainers, element can be removed without disconnecting filter body from the line.

Figure 13.15- Disassembling Pump Strainer (Courtesy of Womack) 33

33

- **Replacement:** A strainer should be replaced in case if :
  - There are holes in the mesh or has physical damage.
  - There are heavy spots of varnish.
  - Unwastable strainers.

Heavily Varnished Strainer   Damaged Suction Filter

Figure 13.16- Pump Strainer that Should be Replaced

34

34

- **Routine:** Should be cleaned routinely, no matter if it looks clean or dirty.

- **Air Blow:** Blowing air from the inside out.

- **If Varnish Found:** Wash it in a solvent + scrubbing with a bristle brush.

- **Solvent:** It should be compatible with the fluid in the system.
  - Mineral oils For example → use kerosene for cleaning.
  - Synthetic or fire-resistant fluids → use same fluid for cleaning the strainer.
  - **Caution!** Do not use gasoline, thinner, etc., which are explosive, highly flammable, and contain no lubricant.

- **Re-installing:** Before reinstalling the strainer, inspect all joints in the inlet plumbing for air leaks, particularly at union joints.

35

35

❖ **Step 2: Pump and Relief Valve (Refer to Fig. 13.14):**

If step 1 didn't solve the problem → :

- Isolate the pump and relief valve from the rest of the system by disconnecting the plumbing (at Point B).
- Fully open the relief valve & run the pump at the right speed.
- Gradually close the relief valve, observe P (at Point A) and Q (at Point C).
- If (P ↑ + Q through the PRV ≈ constant) → pump & PRV are good.
- If (P ↑ + Q through the PRV remarkably decreases) → Pump is bad.
- If (P isn't increasing + Q through the PRV ≈ constant) → PRV is bad.

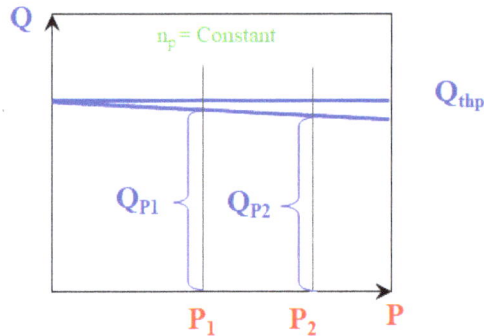

Figure 13.17- Flow-Pressure Characteristics of a Fixed Displacement Pump   36

36

❖ **Step 3: Pump:**

If step 2 shows that the pump is at fault → :

- Check for power take-off from the prime mover, slipping belts, sheared shaft key or pin, broken shaft, broken coupling.

- Check for high working temperature increases the internal leakage.

- Disassemble the pump and inspect the internal parts for wear.

- Perform detailed inspection by applying the chart:
- **"T-Pump-02: Low Flow out of the Pump".**

37

❖ **Step 4: Pressure Relief Valve:**
If step 2 shows that the PRV is at fault → :

- For a quick proof, replace PRV with a new one and repeat the test process.

- Check the valve body and its connection to a subplate or pipes.

- Disassemble the valve, check for broken element or weakened spring.

- Check also for free movement of the spool or poppet.

- Blow air in the orifice of pilot-operated relief valves.

- Clean the valve and assemble it in the system.

- Perform detailed inspection by applying the chart:
- **"T-Valve-03: PCV Troubleshooting".**

38

38

❖ **Step 5: Cylinder:**
If the tested components so far were found working fine → Test Cylinder:
- **External Leakage:**
  o Small leakage is observed as drops of oil and accumulated dirt around it.
  o Excessive leakage is observed as a continuous stream of oil leaking out of the cylinder.

Small Leakage   Excessive Leakage

Figure 13.18- External Leakage in Hydraulic Cylinders

39

39

- **Internal Leakage:**
- Leakage across piston seals is not as easy to detect → slow the cylinder and drifts the load.

- Test Cylinder for Internal Leakage:
  o Fully retract the cylinder and apply full pressure on the rod side.
  o Observe the leakage from the piston side cylinder port.
  o Fully extend the cylinder.
  o Repeat the test to check leakage across the piston from the other side.
- Perform detailed inspection by applying
- **"T-Cylinder-01: Cylinder Troubleshooting".**

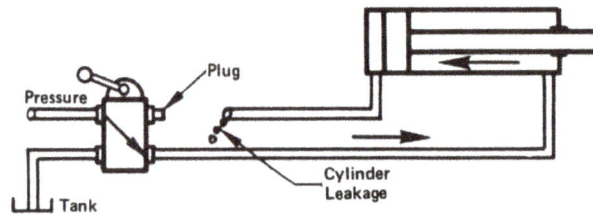

**Figure 13.19- Internal Leakage Test of Hydraulic Cylinders
(Courtesy of Womack)**

40

40

- ❖ **Step 6: Directional Control Valve:**
  If the tested components so far were found working fine → Test DCV:

- For a quick proof, replace DCV with a new one and repeat the test process.

- Check for badly worn spool.

- Check for spool leakage.
  o Disconnect the tank return line.
  o Observe leakage flow through the tank port of the valve.
- Perform detailed inspection by applying chart
- **"T-Valve-01: DCV Troubleshooting".**

41

41

13.11- Troubleshooting of Closed Hydraulic Circuit (Hydrostatic Transmission)

Before going to troubleshoot, the following should be prepared:

- A copy of the factory service manual for the machine to be serviced.
- 400-700 bar (6000 – 10000 psi) high-pressure gauge.
- 35 bar (500 psi) low-pressure gauge.
- Port adaptors for connecting the gauges.
- Short lengths of high-pressure and low-pressure hose and fittings.
- Usual mechanics tools.

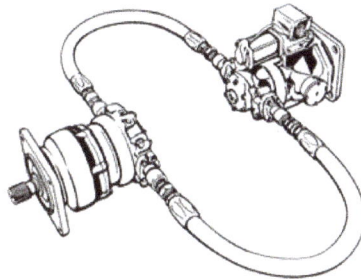

Figure 13.20- Hydrostatic Transmission (Courtesy of Womack)

42

42

Typical hydrostatic transmission circuit diagram isn't that simple.

1. Charge pump.
2. Check valves.
3. Pressure relief valve for the charge pump.
4. Low pressure side hot oil shuttle valve.
5. Hot oil relief valves.
6. Overload pressure relief valves for the two sides of the hydrostatic transmission.

Video 332 (7 min)

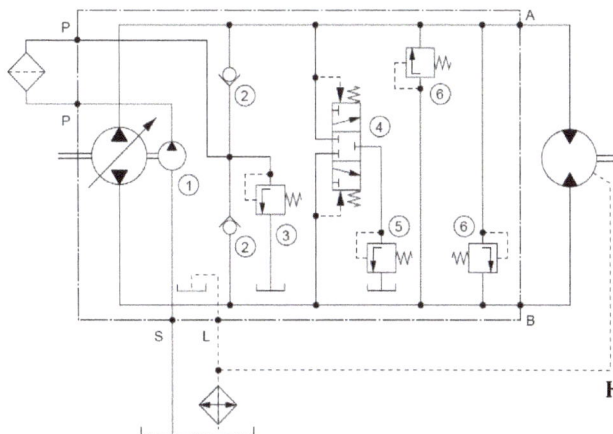

Figure 13.21- Typical Circuit Diagram of Hydrostatic Transmission

43

43

Be reminded to avoid the following common mistakes:

- DO NOT screw taper pipe threads into straight thread. Most transmissions have no taper pipe threads on them.

- DO NOT ever plug the case drain.

- DO NOT change the low-pressure relief valve setting on transmissions which were previously running normally.

- DO NOT change the setting of the high-pressure relief valves unless you have the instruments and the factory instruction manual showing how to re-set them. They must be set to about 500 PSI > than the pressure compensator setting in the manner outlined in the manual.

Video 333 (9 min)

44

44

T-System-09-Troubleshooting of Closed Hydraulic Circuits (Hydrostatic Transmission)	
Prechecks:	<ul><li>Whenever a problem occurs, always check first:</li><li>Oil levels.</li><li>Power transmission to the main pump for possible damage, e.g. broken shafts or couplings, slipping belts, etc.</li><li>Control linkages.</li><li>Gaskets and seals.</li><li>Inspect the high-pressure hoses or pipes between the pump and motor and replace any suspected lines.</li></ul>
System is excessively hot?	<ul><li>Consult Chart:</li><li>**"T-System-04-Excessive System Heat"**.</li></ul>
System response is sluggish?	<ul><li>Check setting of main pump control.</li><li>Check charge pump low pressure and low-pressure PRV (See Note 1).</li></ul>

Table 13.11- Troubleshooting Chart
(T-System-09-Troubleshooting of Closed Hydraulic Circuits)

45

45

Transmission operates in one direction only, or problem shows up only in forward or reverse motion?	▪ Check main pump control. ▪ Check leaking, sticking, and setting of the PRV for the nonworking side of the transmission. ▪ Check Shuttle Valve.
▪ Loss of power and/or speed in either direction? ▪ Motor may run when unloaded but will not produce full torque or speed.	▪ Check engine for correct no-load RPM. ▪ Run engine at load and check for proper performance. ▪ Check fluid level in the reservoir. ▪ Check inlet filter. ▪ Check low pressure of charge pump **(See Note 1)**. ▪ Check high pressure of main pump **(See Note 2)**.
Excessive Case Drain **(See Note 3)**.	Check case drain considering the following: ▪ Check for the pump and the motor separately. ▪ Combined case drain won't give indication which component is at fault. ▪ Check case drain when pump displacement increases. ▪ Check case drain when load is applied to the system. ▪ Compare measured case drain with the rated values. ▪ Excessive case drain flow indicates serious problem in the pump.

Table 13.11- Continue

46

46

Note 1: Checking Low Pressure of Charge Pump:

❖ Complete loss of charge pump pressure at any percentage of the main pump displacement is due to:
- Broken drive shaft or coupling to the charge pump.
- Spring breakage, damage, or dirt in the low-pressure relief valve.

❖ Fluctuating of charge pump pressure at any percentage of the main pump displacement, is due to:
- Cavitation of the charge pump.
- Low oil level in the reservoir.
- Collapsed suction hose.
- Dirty inlet filter.

❖ Charge pump pressure drops only when pump displacement increases is due to:
- Worn charge pump is worn out and experiences large internal leakage.

47

47

Note 2: Checking High Pressure of Main Pump (excerpted from Womack):

- **Step 1: Break the Machine:**
  o Block hydraulic motor shaft by breaking the machine or blocking drive wheels.

- **Step 2: High Pressure Obtained:**
  o Increase pump displacement (forward or reverse).

  o If the vehicle brakes are set, pressure should immediately pick up to the value required by the load.

  o If the vehicle brakes are set, the pressure would immediately pick up to the maximum pressure setting by each side's PRV or pump compensator setting.

48

48

- **Step 3: High Pressure Can't Be Developed:** If there is little or no pressure rise in the system, check the following:

  o **High-pressure relief valves** may be stuck or damaged.

  o **Charge pump** isn't properly operative (follow instructions in Note 1).

  o **Power transmission to main pump** may possibly is damaged.

  o **One or both check valves:** likely one check valve is at fault at a time. Motor will not build up torque in either direction unless both check valves unlikely at fault at the same time.

  o **Leaking shuttle valve** from high pressure side to low pressure side. This could occur due to excessively worn spool.

  o **Main pump controller (pressure compensator)** internal parts may be jammed, dirty, or damaged. If parts of the compensator are removed, it must be re-set to its original setting by service manual procedure.

  o **Excessive case drain** indicates worn pump or motor.

49

49

Note 3: Checking Case Drain:

- Tolerances between the pistons and barrel ≈ 0.0004 inch →
- Normal case drain flow rate ≈ (1 to 3)% of Qpump.
- For example, a 30 (GPM) → case drain flow rate ≈ (0.3 to 0.9) GPM
- Severe increase in case drain flow → oil temperature to rise considerably.
- If case drain flow rate >= 10% of Qpump → pump must be changed.

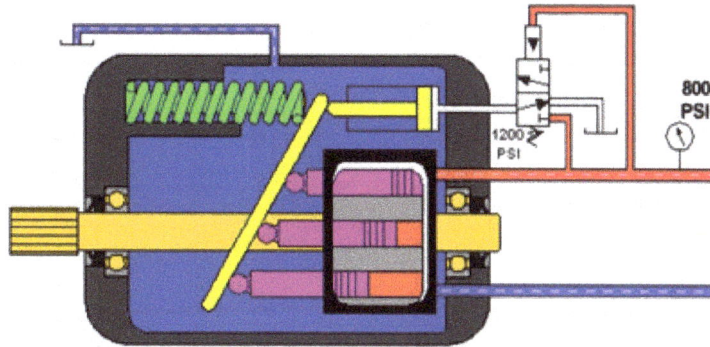

**Figure 13.22- Variable-Displacement, Pressure-Compensated Pump
(Courtesy of Noria)**

50

50

13.12- Actuator Slow Performance

- Charts **T-Actuator-01** through **T-Actuator-07** are applicable for both cylinders and motors.

- If the fault isn't identified by then → apply the following charts if needed: **T-Motors-01 and T-Cylinder-01** in Chapters 5 and 6; respectively.

- Actuator performs slow →
- o Reduced productivity.
- o Increased energy consumption.
- o Automated machines may go out of sequence.

51

51

T-System-10-Actuator Slow Performance	
Hydraulic Fluid: Fluid is aerated?	▪ Consult Chart: ▪ **"T-System-01-Fluid Aeration".**
Oil viscosity is too high?	▪ Check oil viscosity and if the machine is too cold.
Pump: Low flow out of the pump?	▪ Consult Chart: **"T-Pump-02-Low Flow out of the Pump".**
Variable displacement pump was set improperly?	▪ Check and reset to the specified value.
Working Temperature: ▪ System is excessively hot?	▪ Consult Chart: ▪ **"T-System-04-Excessive System Heat".**

Table 13.12- Troubleshooting Chart
(T-System-10-Atuator Slow Performance)

52

52

Accumulator: Accumulator used to boost cylinder speed	▪ Consult Chart: **"T-Accumulator-01-Accumulator Troubleshooting".**
FCV: Flow control valve was set improperly?	▪ Check and reset the valve.
PRV: Pressure relief valve was set too low?	▪ Check and reset the valve.
DCV: Internal leakage in the directional valve?	▪ Check the suspected valve.
Insufficient pilot pressure makes the DCV's spool shifted partially?	▪ Check the pilot pressure of the valve.
Transmission Lines: Restricted pressure lines?	▪ Replace the line.

Table 13.12

53

53

13.13- Actuator Fast Performance

- Actuator performs fast →
- Increased energy consumption.
- Automated machines may go out of sequence.

T-System-11-Actuator Fast Performance	
Pump: Excessive flow out of the pump?	• Consult Chart: • **"T-Pump-04-Excessive Flow out of the Pump".**
Actuator: Actuator undersized?	• Check and resolve accordingly.
Actuator is a variable displacement motor?	• Check the motor size adjustment. • Check proper motor controller operation.
Actuator speed is automatically controlled?	• Check control system settings. • Check proper speed sensor operation.
Load: Overrunning load?	• Check external load conditions. • Check applied method and relevant valve for controlling overrunning load **(See Note 1).**

Table 13.13- Troubleshooting Chart
(T-System-11-Atuator Fast Performance)

54

54

Note 1:
- Each method has pros and cons (review volume 1).
- However, adjustment of the counterbalance valve or the throttle-check valve may change the speed of the load.

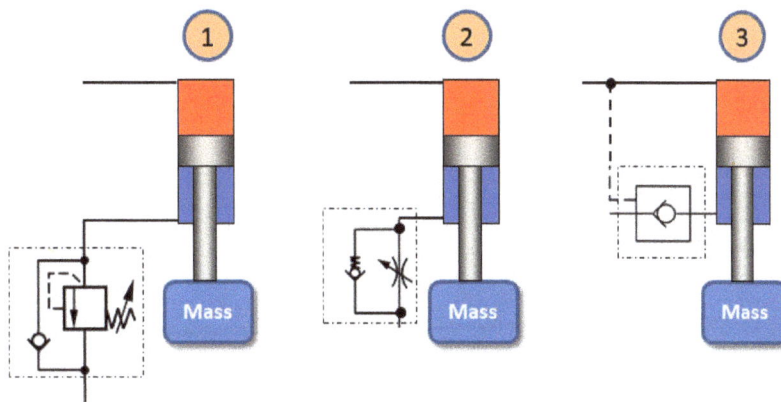

Fig. 13.23- Controlling Overrunning Load

55

55

13.14- Actuator Erratic Performance

- Actuator performs Erratically →
- o Severe vibration, high pitched noise or chatter.
- o Jerky motion and shaking the load that is attached to the actuator.
- o Risk of injuries in mobile machines.

Fig. 13.24- Actuator Erratic Performance
(Courtesy of Fluid Power Safety Institute)

56

56

T-System-12-Actuator Erratic Performance	
Pump: Erratic flow out of the pump?	▪ Consult Chart: ▪ **"T-Pump-03-Erratic Flow at the Pump Outlet".**
Erratic pressure at the pump outlet?	▪ Consult Chart: ▪ **"T-Pump-07-Erratic Pressure at the Pump Outlet".**
Actuator: Leaking actuator?	▪ Consult Chart: ▪ **"T-System-16-Actuator Leaks".**
Air in actuator is not bled properly?	▪ Bleed air from actuator, review manufacturer's instructions.
Cylinder Stick-Slip motion	▪ Check piston and rod seals condition **(Note 1)**. ▪ Check if there is radial force acting on the rod.
Motor spinning below the rated speed?	▪ Review the motor working speed versus the minimum allowable speed. ▪ Adjust the motor speed accordingly.

Table13.14- Troubleshooting Chart
(T-System-12-Atuator Erratic Performance)

57

57

| Load:
Increased friction of outside load?	▪ Check excessive side loading.
Lack of lubrication/greasing to the machine joints and linkages?	▪ Lubricate/grease the joints and linkages.
DCV:
Pressure spikes when switching a directional valve?	▪ Unnecessary long hoses used.
▪ Valve switches too fast **(Notes 2).**
Actuator movement based on shifting a directional valve or an EH valve?	▪ Consult Chart:
▪ **"T-Valve-01- DCV Troubleshooting".**
▪ **"T-Valve-04- EH Valve Troubleshooting".** |
| System Design:
Actuator movement based on a pump connected in parallel to an accumulator? | Check if the pump switches on and off too often because **(Notes 3).**
▪ Improper sizing of pump-accumulator combination.
▪ Improper setting of accumulator charge valve or pressure switch. |

Table 13.14 - Continue

58

58

Note 1 (Stick-Slip):
- The spontaneous *jerking* motion that can occur while two objects are sliding over each other."
- In cylinders, seals are often thought to be the source of the stick-slip.

Fig. 13.25- Deteriorated Seal causes Stick-Slip Cylinder Motion (Courtesy of Parker)

59

59

Note 2 (Pressure Spikes due to Switching DCV):
- Review Volume 1 for construction and operation.
- Proper adjustment of Switching Time Adjustor→ reduced pressure spikes.

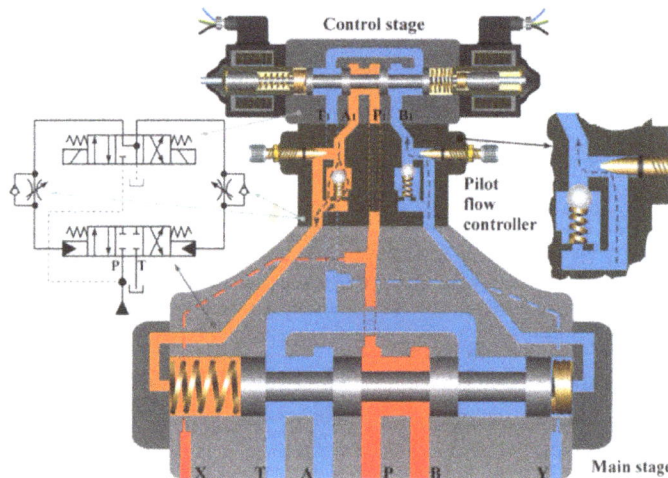

Fig. 13.26 - Pilot-Operated 4/3 Spool-Type Solenoid-Actuated DCV (Courtesy of ASSOFLUID)

60

60

Notes 3 (pump is loaded/unladed very often):
- Pump and accumulator sizes aren't compromised
- Or narrow pressure threshold of the pressure switch
- → the pump will be loaded/unloaded more frequently,
- → the cylinder could move erratically.

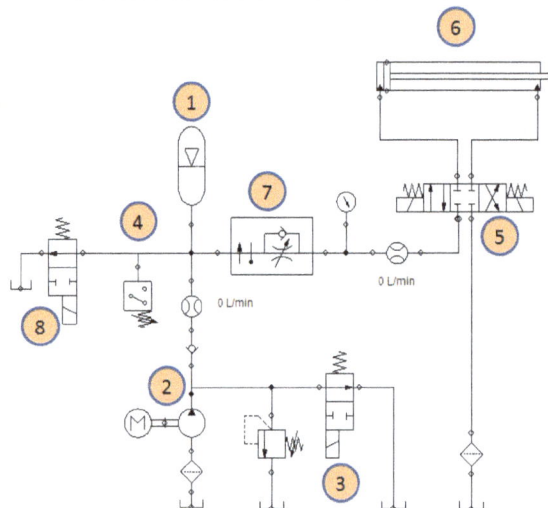

Fig. 13.27-Energy Storage Application Example

61

61

13.15- Actuator Moves in Wrong Direction

- Actuator performs Erratically →
- Automated machines may go out of sequence.
- Possible machine destruction.
- Risk of personal injury due to unexpected machine movements.

T-System-13-Actuator Moves in Wrong Direction	
Pipework: Wrong piping connections with the directional valve or the actuator?	▪ Check and resolve accordingly.
DCV: Directional Valve replaced recently?	Check: ▪ Ordering code. ▪ Proper assembly of valve spool.
Wrong wiring of solenoid-operated directional valve?	▪ Check and resolve accordingly.
Faulty operation of a directional valve	▪ Consult Chart: ▪ "T-Valve-01-DCV Troubleshooting".
Control System: Wrong signal generation from the control system?	▪ Check and resolve accordingly.

Table 13.15- Troubleshooting Chart (T-System-13-Atuator Moves in Wrong Directions) 62

62

13.16- Actuator Stops to Move

T-System-14-Actuator Stops to Move	
Pump: No flow out of the pump?	▪ Consult Chart: ▪ "T-Pump-01-No Flow out of the Pump".
No pressure at the pump outlet?	▪ Consult Chart: ▪ "T-Pump-05-No Pressure at the Pump Outlet".
System Design: Free recirculation of oil to reservoir being allowed through system? **(See Note1).**	▪ Review the sequence of the operation. ▪ Check the proper operation of open-center and tandem-center valves.
Load: Load is mechanically braked or blocked?	▪ Check braking system if found. ▪ Uncouple the actuator from load and check the operation of the actuator.

📹 Video 196 (0.5 min)

**Table 13.16- Troubleshooting Chart
(T-System-14-Atuator Stops to Move)** 63

63

Actuator: Actuator moves when the load is reduced?	• Consult Chart: • "T-System-08-Low Power System".
Actuator undersized and relief valve open?	• Review the size of the actuator or maximum pressure setting.
Actuator broken?	• Repair if possible or replace.
Actuator moves based on operation of a sequence valve	• Check if the sequence valve was set too high.
Actuator moves based on operation of a check valve	• Check if the check valve is jammed. • Check if the pilot operated check valve is jammed or not receiving pilot pressure.
Actuator moves based on shifting a directional or an EH valve?	• Consult Chart: • "T-Valve-01- DCV Troubleshooting". • "T-Valve-04- EH Valve Troubleshooting".

Table 13.16- Continue

64

64

Note (1):
- A pump drives multiple actuators in parallel,
- & one of the DCV valves has a tandem or open center
- → pressure can't be developed in the system.

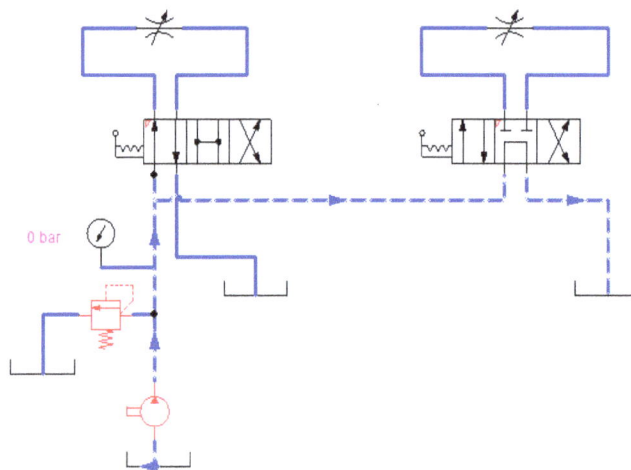

Fig. 13.28- Improper Combination of Tandem and Open Center Valves in Parallel

65

65

13.17- Actuator Load Drifts

Handling heavy loads:
- Presents particular challenges.
- Demands the highest level of precision and safety.

Fig. 13.29- Heavy Lift Crane (Courtesy from Liebherr)

T-System-15-Actuator Load Drifts	
Actuator: Leaking actuator?	• Consult Chart: • **"T-System-16-Actuator Leaks".**
Load: Failure of external mechanical braking system?	• Check and act accordingly.
DCV: Null position of directional valve is biased OR directional valve leaks at null?	• Consult Chart: • **"T-Valve-01-DCV Troubleshooting".**

Table 13.17- Troubleshooting Chart (T-System-15-Atuator Load Drifts)

66

66

13.18- Actuator Leaks
- Leaking Actuator →
- o Missing expensive oil.
- o Environmental pollution and increasing cost of removing it.
- o Possible actuator loss of power and load failure.

Video 355 (8.5 min)

T-System-16-Actuator Leaks	
Cylinder rod scored, scratched or bent?	• Check and replace rod if needed.
Improper torqueing of tie rods cylinder or motor housing?	• Re-torque per manufacturer specifications.
Motor's shaft or cylinder rod subject to radial forces?	• Check maximum allowable radial force. • Review the actuator-load attachment.
Piston seal failure (see **Note 1**)?	• Consult Chart: • **"T-Seal-01-Seal Troubleshooting".**

Table 13.18- Troubleshooting Chart (T-System-16-Atuator Leaks)

67

Note 1:

Piston seals are not leak-tight → the cylinder drifts under external force.

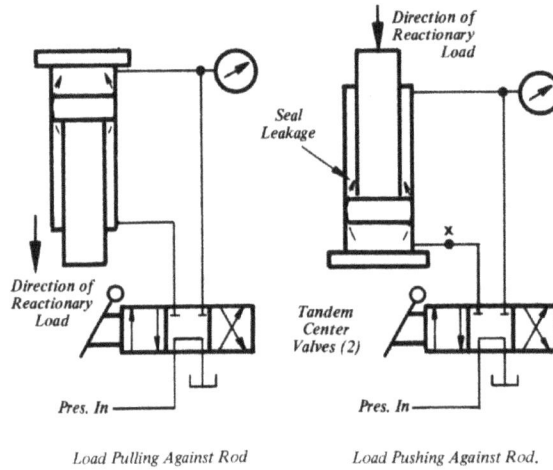

**Fig. 13.30- Cylinder Drift due to Piston Seal Leaking
(Courtesy from Womack)**

68

Chapter 13 Reviews

1. When hydraulic fluid is aerated, this is likely because of?
    A. Air leak into the pump suction.
    B. Suction strainer is clogged.
    C. Pump-motor coupling misalignment.
    D. All of the above.

2. When cavitation is developed at the pump suction, this is likely because of?
    A. Air leak into the pump suction.
    B. Suction strainer is clogged.
    C. Pump-motor coupling misalignment.
    D. All of the above

3. When a hydraulic system experiences excessive noise and vibration, this is likely because of?
    A. Air leak into the pump suction.
    B. Suction strainer is clogged.
    C. Pump-motor coupling misalignment.
    D. All of the above

4. Inspecting a hydraulic system, it was found very hot. Possible consequences of that is?
    A. Actuators move faster.
    B. Reduced fluid viscosity, lack of lubrication, increased internal leakage,
    C. Working pressure is increased.
    D. None of the above.

5. When a hydraulic-driven mobile machine experiences low power, other than the hydraulic system, likely this is because?
    A. Leak in hydraulic actuators.
    B. Relief valve is set too low.
    C. Insufficient power and torque at the power tale-off.
    D. All of the above.

6. An electro-hydraulic position control system went out of sequence. Which of the following component should be listed among the suspicious components to be checked?
    A. Drive motor of the pump.
    B. Suction filter.
    C. Position proximity sensors.
    D. All of the above.

7. When a hydrostatic transmission system responds sluggishly, which of the following needs to be checked?
    A. Accumulator used for riding control.
    B. Steering valve.
    C. Charge pump low pressure and setting of low-pressure relief valve.
    D. Replenishing valve.

8. A recently installed cylinder is installed on an excavator. The cylinder is moving erratically This is likely because of?
    A. Air in the cylinder isn't bled properly.
    B. High flow out of the pump.
    C. Meter-out speed control valve is left open.
    D. Main relief valve is set too high.

9. Optimum range of working temperature when using petroleum-based fluids is?
    A. 20°C to 38°C
    B. 38°C to 54°C
    C. 54°C to 70°C
    D. 70°C to 90°C

10. By OSHA Standard (Act of 1970) the maximum allowable sound level for 8 hours work shift is?
    A. 110 db.
    B. 100 db.
    C. 90 db.
    D. 80 db.

Chapter 13 Assignment

Student Name: -- Student ID: ------------------

Date: -- Score: -----------------------

Question: List the consequences of external leakage from a hydraulic system.

Chapter 14
Examples of Hydraulic Systems Troubleshooting

Objectives:

In this chapter several case studies are presented as examples of applying the logic trouble shooting methodology for hydraulic systems fault detection. In addition, troubleshooting case studies following analytical fault detection methodology are presented. Examples were chosen from both industrial and mobile applications.

Brief Contents:

0

Brief Contents:

14.1-Case Studies Using Logic Fault Detection Methodology

14.2-Industrial Applications Case Studies Using Analytical Fault Detection Methodology

14.3-Mobile Applications Case Studies Using Analytical Fault Detection Methodology

1

14.1-Case Studies Using Logic Fault Detection Methodology

As presented in Chapter 1, steps of logic fault detection procedure are:

- ❖ **Step 1: Review Safety Instructions:**
- ❖ **Step 2: Review Machine History:**
- ❖ **Step 3: Identify Main System Fault:**
- ❖ **Step 4: Apply the System-Level Troubleshooting Chart:**
- ❖ **Step 5: List Suspicious Components:**
- ❖ **Step 6: Perform Preliminary Check on Suspicious Components:**
- ❖ **Step 7: Apply Detailed Check on Suspicious Components:**
- ❖ **Step 8: Fault Evaluation Decision for Repair or Replacement:**
- ❖ **Step 9: Startup and Testing:**
- ❖ **Step 10: Future Considerations and Documentation:**

2

2

14.1.1- Slow Actuator on a Manufacturing Machine

Problem Description:
In a **hydraulic driven manufacturing machine**, a slow actuator was reported by the machine operator.

Fig. 14.1 – Hydraulic Driven Manufacturing Machine

- ❖ **Step 1: Review Safety Instructions:**
- ▪ Safety instructions of the machine/work environment were reviewed.

- ❖ **Step 2: Review Machine History:**
- ▪ Reviewing machine history and circuit diagram →
- ▪ the machine was working fine →
- ▪ unexpectedly the actuator started slowing down and machine productivity is reduced.

3

3

❖ **Step 3: Identify Main System Fault:**   Video 203 (3 min)
- Slow Actuator Performance.

❖ **Step 4: Apply the System-Level Troubleshooting Chart:**
- "T-System-10: Actuator Slow Performance". →
- Aeration in system is identified →
- "T-System-01-Fluid Aeration Fluid Aeration" →
- "T-Pump-11-Air Leaks into the Pump" →
- Leaking fitting on suction line was identified.

❖ **Step 5: List Suspicious Components:**
- SKIP.

❖ **Step 6: Perform Preliminary Check of the Suspicious Components:**
- Skip.

❖ **Step 7: Apply Detailed Check of the Suspicious Components:**
- Skip.

4

4

❖ **Step 8: Fault Evaluation Decision for Repair or Replacement:**
- Tighten the leaking fitting on intake line.
- Replace the pump shaft seal.

❖ **Step 9: Startup and Testing:**
- Completed

❖ **Step 10: Future Considerations and Documentation:**
- Discuss and resolve reasons why fittings on intake line become loose over the time.

- Add the step of checking this fitting among the routine maintenance schedule.

5

5

14.1.2- Interrupted Duty Cycle of a Manufacturing Machine

Problem Description:

- A hydraulic driven manufacturing machine performs a certain duty cycle, in part of which a hydraulic cylinder is reciprocated few times.

- The machine works fine for few days before the duty cycle was interrupted and so that the operator trips the machine and ask for a repair.

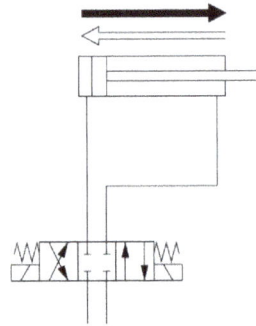

Fig 14.2 – Hydraulic Circuit Diagram of the Investigated Case

6

6

- ❖ **Step 1: Review Safety Instructions:**
- Safety instructions for the machine/work environment were reviewed.

- ❖ **Step 2: Review Machine History:**
- Reviewing machine history and circuit diagram →
- A PLC is used to generate the required sequence of signals to meet the requirement of the duty cycle →
- Duty cycle was recently modified to increase the machine productivity.
- No parts were replaced.

- ❖ **Step 3: Identify Main System Fault:**
- Actuator Stop to Move.

- ❖ **Step 4: Apply the System-Level Troubleshooting Chart:**
- "T-System-14: Actuator Stops to Move" →
- None of the listed reasons were identified.

- ❖ **Step 5: List Suspicious Components:**
- DCV that reciprocates the cylinder.

7

7

❖ **Step 6: Perform Preliminary Check of the Suspicious Components:**
- DCV valve was inspected using the relevant inspection sheet (Table 14.1).
- EH Directional Control valve Data Sheet was reviewed (Fig. 14.3).
- Start with "T-Unit-01-General Check" → No fault was identified

❖ **Step 7: Apply Detailed Check of the Suspicious Components:**
- "T-Valve-04- EH Valve Troubleshooting" → Valve spool isn't moving → solenoid is burn out.

❖ **Step 8: Fault Evaluation Decision of Repair or Replacement:**
- Burn out solenoid is replaced → same problem occurred after few days.
- It was noted that valve has a switching rate of 7200/hr = 2 Hz.
- The modified control code was found to increase the duty cycle to 4 Hz.
- Solution: Replace the current valve by an equivalent DC-Driven EH-valve that has higher switching frequency.

8

8

Hydraulic Valve Inspection Sheet	
Manufacturer	Bosch Rexroth
Model #	4WE 6 E6X
Serial #	NA
Location	NA
Pressure Control Valve Type	▪ Direct [☐Relief ☐Counterbalance ☐Sequence ☐Reducing] ▪ Pilot _[☐Unloading ☐Over-Center ☐ Motor Brake]
Directional Control Valve Type	▪ # Ports (_4_____) # Positions (_3_____) ▪ Initial/Central Position: (Closed-Center) ▪ Reset [☐Spring ☐Detent] ▪ Actuation: [☐Manual ☐Mechanical ☐Pilot ☐Electrical] ▪ More info (_____)
Flow Control Valve Type	☐Throttle ☐Regulator
EH Valve	Type: [☐ON/OFF ☐Proportional ☐Servo] Signal: (_____) Current = Voltage = 110 AC Power:
Valve Configuration	Operation: [☐Direct "Single-Stage" ☐ Pilot "Multiple stages"] Control: _[☐Direct "Internal" ☐ Pilot "External"] Drain: _[☐ Internal ☐ External] Built-in Check Valve _[☐ Yes ☐ No]
Moving Element:	☐ Poppet Type ☐ Spool Type [☐Linear ☐Rotary]
Mounting	☐ Subplate ☐ Line ☐ Manifold "Screw-In" ☐ Sandwich ☐ Other:
Ports/Flow	Port size = Rated flow Rate =
Conditions	Parts: Seals:
Other Notes:	Table 14.1 – Hydraulic Valves Inspection Sheet

9

9

237/252

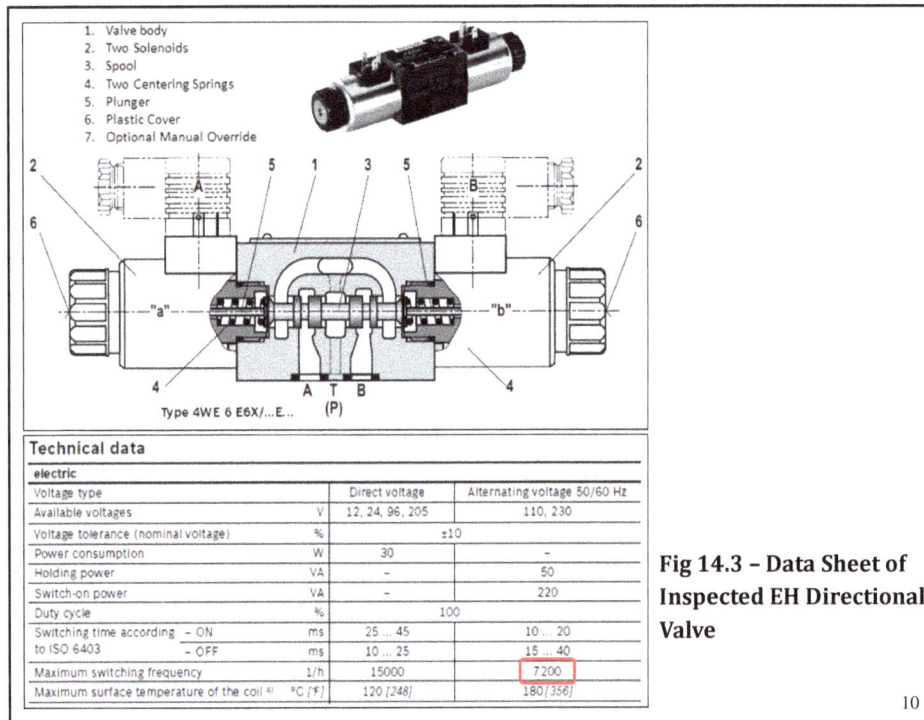

1. Valve body
2. Two Solenoids
3. Spool
4. Two Centering Springs
5. Plunger
6. Plastic Cover
7. Optional Manual Override

Type 4WE 6 E6X/...E...

Technical data

electric			
Voltage type		Direct voltage	Alternating voltage 50/60 Hz
Available voltages	V	12, 24, 96, 205	110, 230
Voltage tolerance (nominal voltage)	%	±10	
Power consumption	W	30	–
Holding power	VA	–	50
Switch-on power	VA	–	220
Duty cycle	%	100	
Switching time according – ON to ISO 6403	ms	25 ... 45	10 ... 20
– OFF	ms	10 ... 25	15 ... 40
Maximum switching frequency	1/h	15000	7200
Maximum surface temperature of the coil [4]	°C [°F]	120 [248]	180 [356]

Fig 14.3 – Data Sheet of Inspected EH Directional Valve

10

10

❖ **Step 9: Startup and Testing:**
▪ Completed

❖ **Step 10: Future Considerations and Documentation:**
▪ Advice control personnel to review specs of hydraulic components before changing control codes. Basic and EH training is crucial for control personnel.

11

11

14.1.3- A Winch Failed to Move

Problem Description:

A **hydraulic driven winch** is operated remotely using a pilot operated directional EH-valve. The winch works fine before it stops to move.

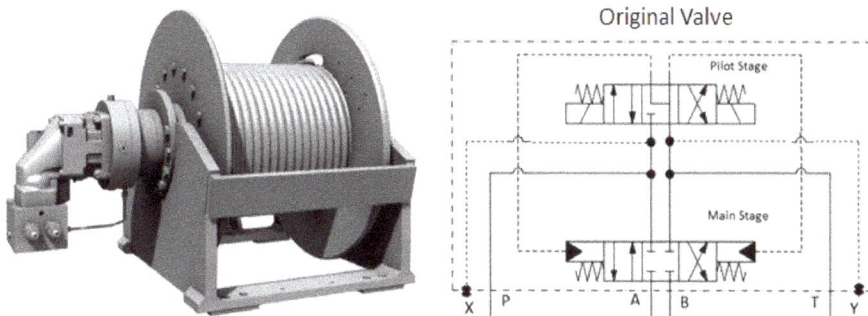

Fig 14.4 – Hydraulic Circuit Diagram of the Investigated Case

12

12

❖ **Step 1: Review Safety Instructions:**

▪ Safety instructions for the machine/work environment were reviewed.

❖ **Step 2: Review Machine History:**

▪ Reviewing machine history and circuit diagram → (Fig. 14.4), it was found that the spool of the Main stage was heavily worn and was replaced by another one.

❖ **Step 3: Identify Main System Fault:**

▪ Actuator Stop to Move.

❖ **Step 4: Apply the System-Level Troubleshooting Chart:**

▪ "T-System-14: Actuator Stops to Move" →

▪ None of the listed reasons were identified.

❖ **Step 5: List Suspicious Components:**

▪ DCV that operates the winch.

13

13

❖ **Step 6: Perform Preliminary Check of the Suspicious Components:**
- DCV valve was inspected using the relevant inspection sheet (Table 14.2).
- EH Directional Control valve Data Sheet was reviewed.
- Start with "T-Unit-01-General Check" → model number was found different
- → Main spool wasn't selected properly (It is noticed that the original valve of a closed-center type is replaced by other spool of an open-center type).

❖ **Step 7: Apply Detailed Check of the Suspicious Components:**
- Skip

❖ **Step 8: Fault Evaluation Decision of Repair or Replacement:**
- If an open-center spool is used in the main stage of a pilot-operated directional valve, one of the following solutions can resolve the problem.
  - ○ Solution 1: Use external (pilot) source to supply the control pressure for the pilot stage.
  - ○ Solution 2: As shown in Fig. 14.5, place a spring-loaded check valve on the pressure port of the main stage.

❖ **Step 9: Startup and Testing:**
- Completed.

14

14

Hydraulic Valve Inspection Sheet	
Manufacturer	Bosch Rexroth
Model #	NA
Serial #	NA
Location	NA
Pressure Control Valve Type	▪ Direct [□Relief □Counterbalance □Sequence □Reducing] ▪ Pilot _[□Unloading □Over-Center □ Motor Brake]
Directional Control Valve Type	▪ # Ports (_ 4) # Positions (3) ▪ Initial/Central Position: (Closed-Center) ▪ Reset [□Spring □Detent] ▪ Actuation: [□Manual □Mechanical □Pilot □Electrical] ▪ More info (_)
Flow Control Valve Type	□Throttle □Regulator
EH Valve	Type: [□ON/OFF □Proportional □Servo] Signal: (_) Current = Voltage = 110 AC Power:
Valve Configuration	Operation: [□Direct "Single-Stage" □ Pilot "2-stages"] Control: _[□Direct "Internal" □ Pilot "External"] Drain: _[□ Internal □ External] Built-in Check Valve _[□ Yes □ No]
Moving Element:	□ Poppet Type □ Spool Type [□Linear □Rotary]
Mounting	□ Subplate □ Line □ Manifold "Screw-In" □ Sandwich □ Other:
Ports/Flow	Port size = Rated flow Rate =
Conditions	Parts: Seals: **Table 14.2 – Hydraulic Valves Inspection**
Other Notes:	Spool of the main stage is open-center type.

15

15

❖ **Step 10: Future Considerations and Documentation:**

- Circuit diagram must be updated as shown in Fig. 14.5.

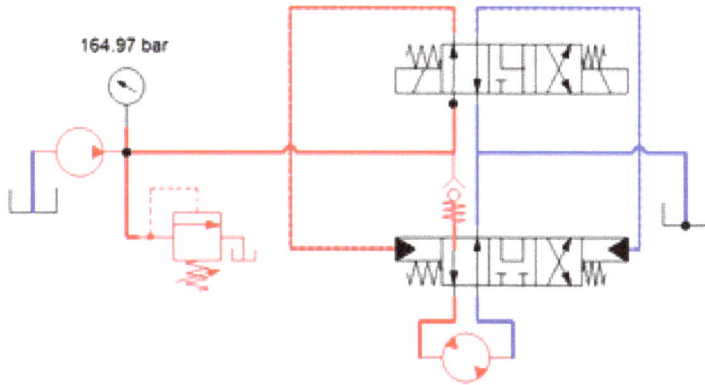

Fig 14.5 – Updated Hydraulic Circuit Diagram to Resolve the Problem

16

16

14.1.4- EH Cylinder Deceleration System Isn't Working Properly

Problem Description:

An **EH cylinder-deceleration system** isn't operating properly. The cylinder does not decelerate at position S2 where it suppose to?

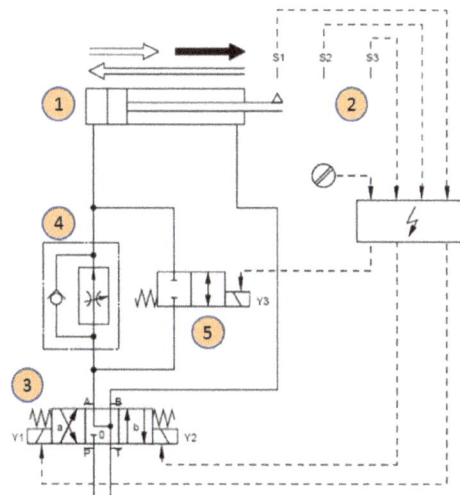

**Fig. 14.6- Electro-Hydraulic Cylinder
Deceleration Circuit
(Courtesy of Bosch Rexroth)**

17

17

❖ **Step 1: Review Safety Instructions:**
- Safety instructions for the machine/work environment were reviewed.

❖ **Step 2: Review Machine History:**
- Reviewing the machine history and circuit diagram →
- No recent changes were reported.

❖ **Step 3: Identify Main System Fault:**
- No recent changes were reported.

❖ **Step 4: Apply the System-Level Troubleshooting Chart:**
- "T-System-06: Faulty System Sequence" → power line of the switch S2 was accidentally disassembled from the mounting terminal.

❖ **Step 5: List Suspicious Components:**
- Skip.

18

18

❖ **Step 6: Perform Preliminary Check of the Suspicious Components:**
- Skip

❖ **Step 7: Apply Detailed Check of the Suspicious Components:**
- Skip

❖ **Step 8: Fault Evaluation Decision for Repair or Replacement:**
- Switch S2 is wired properly.
- Main chassis, on which the control panel is mounted, is subjected to permanent vibration. Control panel is relocated to another fixed frame.

❖ **Step 9: Startup and Testing:**
- Completed

❖ **Step 10: Future Considerations and Documentation:**
- Control panels should be located apart from the machine body if possible. Otherwise, it must be located apart from vibrating chasses.

19

19

14.1.5- Steady State Error in Cylinder Position Control System

Problem Description:

An **EH motor in a speed control system**, is experiencing steady state error. The motor runs slower than the desired reference speed set by the control system.

Fig. 14.7- Electro-Hydraulic Cylinder Position Control System

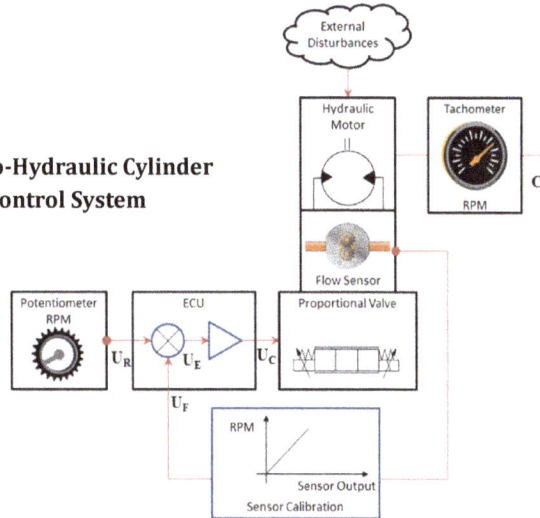

20

20

❖ **Step 1: Review Safety Instructions:**
- Safety instructions for the machine/work environment were reviewed.

❖ **Step 2: Review Machine History:**
- Reviewing machine history → the machine was working fine →
- Steady state error arises over the time → motor slowing down →
- Some calibration was done to remove the steady state error →
- The machine get back to work fine →
- Steady state error starts to appear again after a month of operation.
- Reviewing hydraulic and control circuit diagrams → control loop is closed on the flow rate that drives the motor.

❖ **Step 3: Identify Main System Fault:**
- Actuator slow performance.

❖ **Step 4: Apply the System-Level Troubleshooting Chart:**
- "T-System-10: Actuator Slow Performance" → No fault was identified.

21

❖ **Step 5: List Suspicious Components:**
- Any elements in the system could be suspicious.
- However, control loop is closed on the flow →
- Control system is not able to detect errors resulting from motor →
- Wisely, check the motor first.

❖ **Step 6: Perform Preliminary Check of the Suspicious Components:**
- Motor was inspected using the relevant inspection sheet.
- Motor Data Sheet was reviewed.
- "T-Unit-01-General Check" → No fault was identified.

❖ **Step 7: Apply Detailed Check of the Suspicious Components:**
- "T-Motor-01-Motor Troubleshooting" → "T-Pump-10-Excessive Pump Wear (As applied for Motors)" → High water content in the fluid.

22

22

❖ **Step 8: Fault Evaluation Decision for Repair or Replacement:**
- Fluid contamination with water → loss of lubricity →
- Gradual increase in wear of the internal rotating elements →
- Gradual increase of internal leakage →
- motor gradually slowing down →
- Control system can't detect and compensate the fault because motor wasn't part of the control loop →
- Motor was replaced.
- Since the system is small, it was advised to drain, flush, and change the fluid.
- Source of water penetration into the system was investigated and resolved.

❖ **Step 9: Startup and Testing:**
- Completed.

❖ **Step 10: Future Considerations and Documentation:**
- Frequent fluid analysis is required.

23

23

14.1.6- Loss of Power Accompanied by an Increase of Pump Noise

Problem Description:

- In the shown general hydraulic system →
- Gradual or sudden loss of high pressure → cylinder stall under light loads →
- Loss of cylinder power accompanied by an increase in pump noise.

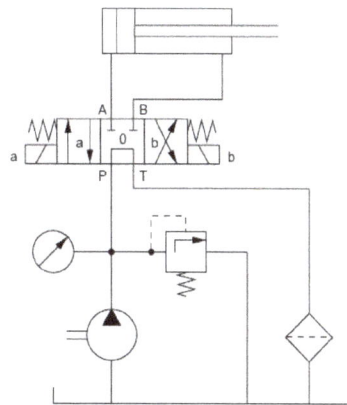

Fig. 14.8- Hydraulic Circuit of the Investigated Case   24

24

❖ **Step 1: Review Safety Instructions:**

- Safety instructions for the machine/work environment were reviewed.

❖ **Step 2: Review Machine History:**

- Reviewing the machine history → hydraulic fluid was recently replaced.

❖ **Step 3: Identify Main System Fault:**

- Pump noise and loss of power when building pressure.

❖ **Step 4: Apply the System-Level Troubleshooting Chart:**

- "T-Pump-12: Excessive Pump Noise & Vibration" →
- "T-System-02: Pump Cavitation" → Fluid is too viscous.

❖ **Step 5: List Suspicious Components:**

- Skip.

25

25

❖ **Step 6: Perform Preliminary Check of the Suspicious Components:**
▪ Skip.

❖ **Step 7: Apply Detailed Check of the Suspicious Components:**
▪ Skip.

❖ **Step 8: Fault Evaluation Decision for Repair or Replacement:**
▪ Replacing the hydraulic fluid with another of higher viscosity →
▪ Highly negative pressure in the suction line.
▪ Fluid was drained and replaced with fluid specified by the machine manufacturer.

❖ **Step 9: Startup and Testing:**
▪ Completed.

❖ **Step 10: Future Considerations and Documentation:**
▪ Adhere to specifications published by the machine manufacturer.

26

26

14.1.7- Mold of an Injection Molding Machine is Partially Filled

Problem Description:
▪ A hydraulic-driven injection molding machine →
▪ the mold is partially filled → batches inconsistency.

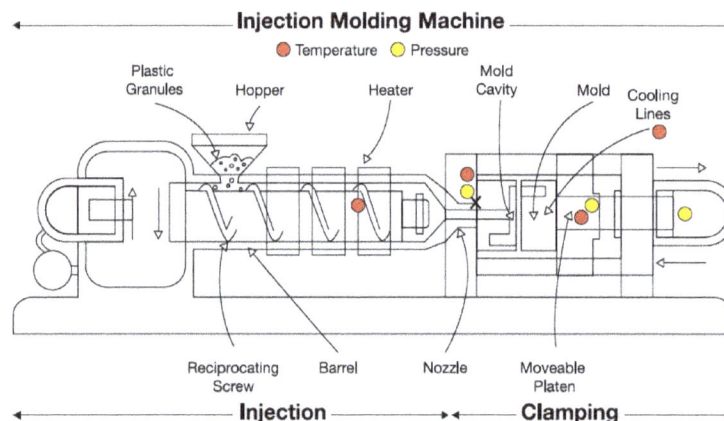

**Fig. 14.9- Hydraulic-Driven Injection Molding Machine
(Courtesy of Parker)**

27

27

❖ **Step 1: Review Safety Instructions:**
- Safety instructions for the machine/work environment were reviewed.

❖ **Step 2: Review Machine History:**
- Reviewing the machine history and service manual →
- No reasons for the indicated symptoms.

❖ **Step 3: Identify Main System Fault:**
loss of injection pressure → mold isn't fully filled is →
Main fault is low power.

❖ **Step 4: Apply the System-Level Troubleshooting Chart:**
- "T-System-5: Low Power System" →
- "T-Pump-06: Low Pressure at the Pump Outlet".
- → Pressure relief is set too low.

❖ **Step 5: List Suspicious Components:**
- Skip.

28

28

❖ **Step 6: Perform Preliminary Check of the Suspicious Components:**
- Skip.

❖ **Step 7: Apply Detailed Check of the Suspicious Components:**
- Skip.

❖ **Step 8: Fault Evaluation Decision for Repair or Replacement:**
- Reset the valve.

❖ **Step 9: Startup and Testing:**
- Completed

❖ **Step 10: Future Considerations and Documentation:**
- Add a wireless sensor → detect the slightest change in pressure →
- Alert users on mobile devices → diagnose fault immediately.

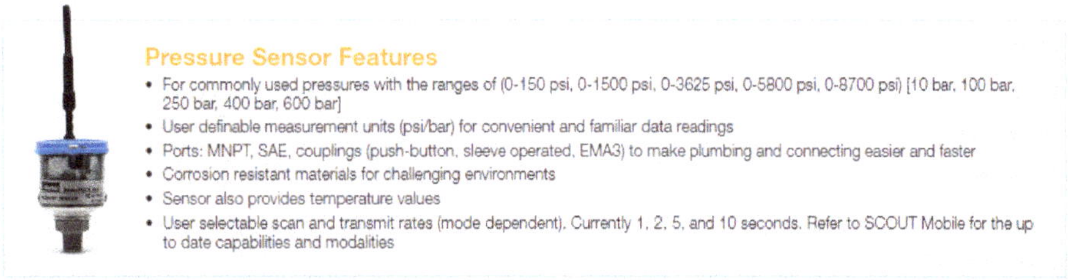

Pressure Sensor Features
- For commonly used pressures with the ranges of (0-150 psi, 0-1500 psi, 0-3625 psi, 0-5800 psi, 0-8700 psi) [10 bar, 100 bar, 250 bar, 400 bar, 600 bar]
- User definable measurement units (psi/bar) for convenient and familiar data readings
- Ports: MNPT, SAE, couplings (push-button, sleeve operated, EMA3) to make plumbing and connecting easier and faster
- Corrosion resistant materials for challenging environments
- Sensor also provides temperature values
- User selectable scan and transmit rates (mode dependent). Currently 1, 2, 5, and 10 seconds. Refer to SCOUT Mobile for the up to date capabilities and modalities

Fig. 14.10- Wireless Pressure Sensor (Courtesy of Parker)   29

29

14.1.8- Excavator Experiencing Low Power

Problem Description:

A trenching excavator that starts to experience slower performance.

❖ **Step 1: Review Safety Instructions:**
- Safety instructions for the machine/work environment were reviewed.

❖ **Step 2: Review Machine History:**
- Reviewing the machine history →
- No recent changes were reported.
- Reviewing the service manual → actual duty cycle is 25% longer in time as compared to the design duty cycle.

Fig 14.11 – Trenching Excavator

❖ **Step 3: Identify Main System Fault:**
- Low power system performance.

30

30

❖ **Step 4: Apply the System-Level Troubleshooting Chart:**
- "T-System-05: Low Power System" →
- Air filter of the engine was found workable in good shape and is not clogged.
- Power transmission elements were found ok, no slippage, worn or broken elements.
- "T-Pump-05: Low Pressure at the Pump Outlet" → no faults were detected.
- "T-Cylinder-01: Cylinder Troubleshooting"→
- "T-Seal-01-Seal Troubleshooting" →
- Piston seals were inspected and found to have axial cuts and particles embedded in the seal material due to abrasive contaminants.

Step 5: List Suspicious Components:
- Skip.

❖ **Step 6: Perform Preliminary Check of the Suspicious Components:**
- Skip.

❖ **Step 7: Apply Detailed Check of the Suspicious Components:**
- Skip.

31

31

❖ **Step 8: Fault Evaluation Decision for Repair or Replacement:**
- Piston seal was replaced.
- System filter cartridge was replaced.
- System was drained, flushed, and filled with clean oil to specs found in the service manual.

❖ **Step 9: Startup and Testing:**
- Completed.

❖ **Step 10: Future Considerations and Documentation:**
- Perform routine maintenance on time and frequent fluid analysis is advised.

32

32

14.2- Case Studies Using Analytical Fault Detection Methodology for Industrial Applications

14.3- Case Studies Using Analytical Fault Detection Methodology for Mobile Applications

This section was graciously provided to this textbook as a Courtesy of "CFC Industrial Training".

Robert Sheaf
President
CFC Industrial Training
A CFC-Solar company
7042 Fairfield Business Dr.
Fairfield, Ohio 45014
513-874-3225

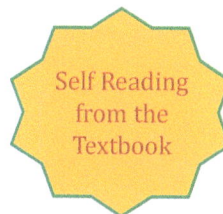

Self Reading from the Textbook

33

33

Additional case studies can be downloaded from:

https://www.compudraulic.com/download-software-for-textbooks

Video 204 (0.5 min)

Video 205 (3 min)

Video 206 (2 min)

Video 207 (0.5 min)

34

34

Answers to Chapters Reviews

Chapter 1:

1	2	3	4	5					
B	D	A	A	B					

Chapter 2:

1	2	3	4	5					
D	B	C	D	A					

Chapter 3:

1	2	3	4	5					
D	A	D	B	D					

Chapter 4:

1	2	3	4	5	6	7	8	9	10
B	C	C	C	D	A	D	C	D	B

Chapter 5:

1	2	3	4	5					
D	A	B	A	D					

Chapter 6:

1	2	3	4	5					
D	D	A	C	B					

Chapter 7:

1	2	3	4	5	6	7	8	9	10
C	B	B	B	A	D	D	C	B	D

Chapter 8:

1	2	3	4	5					
D	B	D	B	A					

Chapter 9:

1	2	3	4	5					
A	B	C	D	A					

Chapter 10:

1	2	3	4	5					
D	C	A	B	D					

Chapter 11:

1	2	3	4	5	6	7	8	9	10
B	C	C	A	B					

Chapter 12:

1	2	3	4	5					
D	D	B	C	A					

Chapter 13:

1	2	3	4	5	6	7	8	9	10
A	B	D	B	D	C	C	A	B	C

www.ingramcontent.com/pod-product-compliance
Lightning Source LLC
Chambersburg PA
CBHW052341210326
41597CB00037B/6211